Boris Nahapetian

Limit Theorems and Some Applications in Statistical Physics

B. G. Teubner Verlagsgesellschaft
Stuttgart · Leipzig 1991

ISBN 978-3-322-93433-8   ISBN 978-3-322-93432-1 (eBook)
DOI 10.1007/978-3-322-93432-1

Recent years have witnessed much progress in the theory of summation
of dependent random variables. The book describes the contemporary
state of several parts of this theory as well as its applications in
the theory of Gibbs random fields, the basicals of which are also
presented.

In den letzten Jahren wurden große Fortschritte in der Theorie der
Summation abhängiger Variabler erzielt.
Dieses Buch beschreibt den gegenwärtigen Stand sowohl der Theorie
als auch ihrer Anwendungen in der Theorie der Gibbsschen zufälli-
gen Felder.

Au cours des derniéres années, de grands progrès ont été réalisés
dans la théorie de la sommation des variables aléatoires dépen-
dentes. Ce livre décrit l'état actuel de certains chapitres de
cette théorie, tout autant que ses applications à la théorie des
champs aléatoires de Gibbs dont on présente les bases.

За последнее время достигнут большой прогресс в теории
суммирования зависимых случайных величин. В книге предпринята
попытка отразить современное состояние некоторых глав этой
теории с применениями к гиббсовским случайным полям, основания
теории которых также приводятся.

During the recent years the interest in limit theorems for dependent random variables has considerably grown being stimulated both by requirements of probability theory and by many applied disciplines. A special place among such disciplines belongs to statistical physics whose mathematical apparatus for a long time has been developing regardless the probability theory. However recently a process of mutual understanding between the specialists of this two fields started to develop. This resulted in a burst of activity for the benefit of both the subjects. For instance, the concept of Gibbs random field which has originated in statistical physics became a powerful stimulus for development of the theory of random fields while the Gibbsian fields themselves turned out to be excellent object for application of many general probabilistic results.
The present book attempts to reflect this development concentrating upon limit theorems for sums of components of weakly dependent random processes and fields.
In the first chapter we formulate some necessary facts from probability theory. The second chapter discusses different weak dependence conditions for random processes and fields. In Chapter 3 the behaviour of the variance of the sum of dependent random variables is investigated and some estimates for moments are given. The last part of this chapter presents several probabilistic inequalities which are the necessary technical tools in our proofs of limit theorems. Chapter 4 is devoted to description of the basic methods of proofs. In particular we present a modification of Bernstein's method. The fifth chapter contains limit theorems for the random processes satisfying the so-called mixing criterions ($\alpha, \varphi, \rho, \psi$ - mixing): Various versions of the central limit theorem (c.l.th.), the invariance principle, the law of the iterated logarithm and rate of convergence in c.l.th. In Chapter 6 we introduce generalized mixing conditions and apply them in some versions of c.l.th. Several resalts from non-commutative probability theory are presented too. The seventh chapter is devoted to limit theorems for random fields. Here again the corresponding limit theorems are given.
Chapter 8 presents known results of Dobrushin about the description oj a random field by means of its conditional distribution. The estimate of mixing coefficient in terms of conditional distribution and the appropriate c.l.th. also are presented. The last Chapter 9 contains elements of the theory of Gibbs random fields with some appli-

cations of the results of the previous chapters.
Note that in this book we as a rule omit the proofs of theorems
which are contained in known monographs /7/, /66/, /70/, /113/,
/114/, /135/.
The material of this book was selected according to the author's
interests and does not reflect all sides of the considered subject.
Additional information one can find in /15/, /17/, /43/, /69/, /81/,
/101/, /103/, /106/, /119/, /132/, /138/.
Finally, I wish to express my sincere thanks to Prof. R.V.Ambart-
zumian who encouraged me to write this text and offered constant
support during concluding phase of the work. I would like to thank
Prof. V.Arzumanian and Prof. S.K.Pogosian as well as the referees
for their helpful advices concerning the mathematics. I am grate-
ful to Prof. S.K.Pogosian and K.P.Kazanchian for their help in im-
proving the English of the original text and to a friendly person
in Teubner Verlag who remained personally unknown to me. These im-
provment could not be incomporated in finally version without ama-
zing technical skill of V.V.Stamboltzian.

Erevan, October,1989.
B.S.Nahapetian.

C O N T E N T S

1. PRELIMINARY PROBABILISTIC RESULTS

In this chapter we give without proof some well known facts of
probability theory which we will need later. The proofs or
references can be found in /7/ , /66/ or /102/ .

1.1 Convergence of Probability Measures

Let $(\Omega, \mathcal{F}, P)$ be a probability space. A measurable
mapping $X : \Omega \to S$ into some measurable space
$(S, \mathcal{B}(S))$ is called a random element. If S is a topo-
logical space, then usually $\mathcal{B}(S)$ is the σ-field of
Borel subsets of S. The complete probabilistic information
about a random element is contained in its distribution, i.e.,
in the probability measure $P(X^{-1}(B))$, $B \in \mathcal{B}(S)$. If
$S \subset R^{k}$ then for $k=1$ X is called a random variable and for
$k > 1$ - random vector. In the case, where S is the space
of real valued functions $x(t)$, $t \in T \subset R^{m}$ the random element
defines a function of two variables $X(t, \omega)$, $t \in T$, $\omega \in \Omega$
which is called the random process when $m = 1$ and the random
field when $m > 1$. Let $\mathcal{B}(S)$ be a cylindrical σ-field,
i.e., a σ-field generated by the following sets

$$\pi^{-1}_{t_{1},\dots,t_{k}}(B), \quad B \in \mathcal{B}(R^{k}), \quad t_{i} \in T, \quad i = \overline{1,k}$$

$$(1.1.1)$$

$$\pi_{t_{1},\dots,t_{k}} x(\cdot) = (x(t_{1}),\dots, x(t_{k})), \quad x(\cdot) \in S .$$

Then by Kolmogorov's theorem the distribution of a random element (field) is defined by the consistent system of finite dimensional distributions

$$P_{t_{1},\dots,t_{k}}(B) = P(\pi^{-1}_{t_{1},\dots,t_{k}}(B)), \quad B \in \mathcal{B}(R^{k}),$$
$$t_{i} \in T, \quad i = \overline{1,k} .$$

It is said that the sequence of probability measures P_{n} converges weakly to the probability measure P ($P_{n} \Rightarrow P$, $n \to \infty$)
if for any P-continuous set $B \in \mathcal{B}(S)$, $P_{n}(B) \to P(B)$ as
$n \to \infty$. The set B is called P-continuous if $P(\partial B) = 0$,
where ∂B is the boundary of the set B , i.e.,the family
of points which are the limit points for the sets B and B'
simultaneously. (B' is the complement of B).
The class of sets $\mathfrak{M} \subset \mathcal{B}(S)$ is called a determining class

if for any two measures P and Q from $P(M) = Q(M)$, $M \in \mathfrak{M}$ it follows that $P = Q$ on $\mathcal{B}(S)$. The class $\mathfrak{M} \subset \mathcal{B}(S)$ is called a convergence-determining class if $P_n \Rightarrow P$, $n \to \infty$ as soon as $P_n(M) \to P(M)$, $n \to \infty$, for any P-continuous $M \in \mathfrak{M}$. It is evident that by virtue of the uniqueness of the limit, the convergence-determining class is a determining one but not inversely.

It is said that the sequence of random elements $X_n, n \in \mathbb{N}$ converges in distribution or weakly to the random element X

$$\left(X_n \xrightarrow{\mathcal{D}} X , \ n \to \infty \right)$$

if the sequence of probability measures $P(X_n^{-1}(B))$, $n \in \mathbb{N}$ converges weakly to the probability measure $P(X^{-1}(B))$, $B \in \mathcal{B}(S)$.

Let us mention a statement connected with some transformations of random elements.

<u>Theorem 1.1.1.</u> Let S , S' be the metric spaces and $h : S \to S'$ be some measurable mapping.

1. If $P_n \Rightarrow P$ as $n \to \infty$ in $(S, \mathcal{B}(S))$ and $P(D_h) = 0$, where D_h is the set of discontinuity points of h then

$$P_n h^{-1} \Rightarrow P h^{-1}, \ n \to \infty \ ;$$

or

2. If $X_n \xrightarrow{\mathcal{D}} X$, $n \to \infty$ and the measurable map h is such that $P_X(D_h) = 0$ then

$$h X_n \xrightarrow{\mathcal{D}} h X, \quad n \to \infty .$$

Now it is clear that if h is a continuous map then $P_n \Rightarrow P$ $n \to \infty$ implies $P_n h^{-1} \Rightarrow P h^{-1}$, $n \to \infty$ also $X_n \xrightarrow{\mathcal{D}} X$, $n \to \infty$ implies $h X_n \xrightarrow{\mathcal{D}} h X$, $n \to \infty$.

The family of probability measures $\mathcal{P} = \{P\}$ is called compact if from any of its subsequences one can extract a weakly convergent subsequence.

<u>Theorem 1.1.2.</u> The sequence of measures P_n , $n \in \mathbb{N}$, has a weak limit if every finite dimensional distribution of the measures P_n has weak limits and the sequence itself is compact.

The family $\mathcal{P} = \{P\}$ is called tight if for every $\varepsilon > 0$ there exists a compact $K_\varepsilon \subset S$ such that

$$P(K_\varepsilon) \geq 1 - \varepsilon \qquad , \text{ for all } \quad P \in \mathcal{P}.$$

Theorem 1.1.3 (Prokhorov). If a family of measures $\mathcal{P}$ is tight then it is compact. If S is a separable and complete space then a compact family of measures is tight.

1.2 Central Limit Theorem for Sums of Random Vectors

Let P be a probability measure on $\mathcal{B}(R^k)$. The function

$$F(x) = P(A(x)), \quad x \in R^k,$$

$$A(x) = \left\{ y \in R^k : -\infty < y_j < x_j, \quad j = \overline{1,k} \right\}$$

is called the distribution function of the measure P. The class of sets $A(x)$, $x \in R^k$ is a measure determining one and the sequence of measures P_n on $\mathcal{B}(R^k)$ converges weakly to the measure P iff $F_n(x) \Rightarrow F(x)$, $n \to \infty$, i.e., $F_n(x) \to F(x)$, $n \to \infty$ in each continuity point of the function $F(x)$. The characteristic function of the measure P (or distribution function $F(x)$) is defined by

$$\varphi(t) = \int_{R^k} e^{i(t,x)} P(dx) = \int_{R^k} e^{i(t,x)} dF(x), \quad x \in R^k,$$

$$(t,x) = \sum_{j=1}^{k} t_j x_j.$$

The characteristic function φ and the corresponding distribution function are mutually determined. Moreover if F_n, $n = 1,2,\ldots$, is a sequence of distribution functions and φ_n, $n = 1,2,\ldots$, is the corresponding sequence of characteristic functions then $F_n \Rightarrow F$ iff $\varphi_n(t) \to \varphi(t)$, $n \to \infty$, for every fixed $t \in R^k$. It is said that the random vector $X = (x_i, \ i = \overline{1,k})$ has a normal distribution with parameters (a, σ^2), if

$$\varphi_X(t) = \exp\left\{ i(a,t) - \tfrac{1}{2}(\sigma^2 t, t) \right\},$$

where $a = EX$, $\sigma^2 = \{ \mathrm{Cov}(x_i, x_j), \ i,j = \overline{1,k} \} = DX$ is the covariance matrix.

Let us consider triangular array of random vectors from R^k

$$X_{nj}, \quad j = \overline{1, K_n}, \quad k_n \to \infty, \quad n \to \infty,$$

where for any n the vectors X_{nj}, $j = \overline{1, k_n}$ are independent.

Let also

$$EX_{nj} = 0, \quad j = \overline{1, k_n} \quad n \in \mathbb{N},$$

$$S_n = \sum_{j=1}^{k_n} X_{nj}, \quad \sigma_n^2 = DS_n, \quad |x| = \left(\sum_{j=1}^{k} x_j^2 \right)^{1/2}, \quad x \in R^k.$$

and $N(0, \sigma^2)$ stands for a Gaussian r.v. with parameters $0, \sigma^2$.

<u>Theorem 1.2.1.</u> Let $\sigma_n^2 \to \sigma^2$ (element-wise) when $n \to \infty$
and for every $\varepsilon > 0$

$$\sum_{j=1}^{k_n} \int_{|X_{nj}| \geq \varepsilon} |X_{nj}|^2 P(d\omega) \to 0, \quad n \to \infty \qquad \text{(Lindeberg condition)}.$$

Then

$$S_n \xrightarrow{\mathcal{D}} N(0, \sigma^2),$$

i.e., the central limit theorem (c.l.th.) is valid.

1.3 Invariance Principle

Let $C[0,1]$ be the space of functions which are continuous in
the uniform metric. Let $P_n \Rightarrow P$, $n \to \infty$, where P_n, $n \in \mathbb{N}$,
and P are probability measures on $\mathcal{B}(C[0,1])$. Then by Theorem 1.1.1 for arbitrary $t_1, \ldots, t_k$ and every $k \in \mathbb{N}$

$$P_n \pi_{t_1, \ldots, t_k}^{-1} \Rightarrow P \pi_{t_1, \ldots, t_k}^{-1}, \quad n \to \infty,$$

in other words, all the finite dimensional distributions converge.
However, the inverse is not true. The class of cylinders is not
a convergence-determining class. Using Theorems 1.1.2 and 1.1.3
we have

<u>Theorem 1.3.1.</u> The sequence of probability measures P_n in
 $C[0,1]$ has a weak limit iff each finite dimensional distribution of the measure P_n has a weak limit and the sequence
itself is tight.

Thus it is important to know when the family of measures is
tight.

Let $\omega(x, \delta) = \sup \left\{ |x(s) - x(t)|, \ |s-t| \leq \delta ; \ t, s \in [0,1] \right\}$
be the continuity module of function $x(t)$, $t \in [0,1]$.

__Theorem 1.3.2.__ A family of measures P_n , $n \in \mathbb{N}$ in $C[0,1]$ is tight iff

1. For any $\varepsilon > 0$ there exists a number $a = a(\varepsilon)$ such that

$$P_n (x(t), t \in [0,1] ; |x(0)| > a) < \varepsilon, \quad n \in \mathbb{N} ;$$

2. For any ε and ε' there exist $\delta, 0 < \delta < 1$ and $n_0 \in \mathbb{N}$ such that

$$P_n (x(t), t \in [0,1] ; \omega(x, \delta) \geqslant \varepsilon) \leqslant \varepsilon', \quad n \geqslant n_0 .$$

__Remark 1.3.1.__ Sometimes instead of Condition 2 of Theorem 1.3.2 it is more convenient to check the following condition:

2') For any positive ε and ε' there exist δ , $0 < \delta < 1$ and $n_0 \in \mathbb{N}$ such that

$$\frac{1}{\delta} P_n (x(t), t \in [0,1] : \sup_{t \leqslant s \leqslant t+\delta} |x(s) - x(t)| \geqslant \varepsilon) \leqslant \eta, \quad n \geqslant n_0 .$$

The random process $W = (\omega(t), t \in [0,1])$ is called Wiener process if

1. $\omega(t), t \in [0,1]$ has independent increments that is for any k and $0 \leqslant t_1 < \ldots < t_k$ the variables $\omega(t_2) - \omega(t_1), \ldots, \omega(t_k) - \omega(t_{k-1})$ are independent;

2. The increments $\omega(s) - \omega(t), 0 \leqslant t \leqslant s$ have the distribution $N(0, s-t)$;

3. $\omega(t), t \in [0,1]$ is a process with continuous a.sh. trajectories and $\omega(0) = 0$ a.sh.

Let $X_n, n \in \mathbb{N}$ be the sequence of centered random variables. Denote $\displaystyle S_n = \sum_{j=1}^{n} X_j$ and put

$$S_n(t) = \frac{1}{6\sqrt{n}} S_{i-1} + n(t - \frac{i-1}{n}) \frac{1}{6\sqrt{n}} X_i , \quad t \in [\frac{i-1}{n}, \frac{i}{n}] ,$$

$$i = \overline{1,n} , \quad 6 > 0 .$$

Let P_n be the distribution of the random element $\{S_n(t), t \in [0,1]\}$. If $P_n \Rightarrow W, n \to \infty$ then we say that the sequence $X_n, n \in \mathbb{N}$ satisfies the invariance principle or functional form of the central limit theorem (f.c.l.th.).

__Theorem 1.3.3.__ The sequence $X_n, n \in \mathbb{N}$ of random variables satisfies f.c.l.th. if the finite dimensional distributions of random elements $\{S_n(t), t \in [0,1]\}$ converge weakly to those of W and for any $\varepsilon > 0$ there exists $\lambda, \lambda > 1$ and an integer n_0 such that

$$P\left(\max_{1\le i\le n}|S_{k+i}-S_k|\ge\lambda\,\sigma\,\sqrt{n}\,\right)\le\frac{\varepsilon}{\lambda^2}\ ,\quad k\in\mathbb{N},\ n\ge n_0\ .$$

As it has been shown by Donsker the i.i.d. sequences of random variables satisfy the invariance principles. Let h be the functional continuous with respect to the uniform metric in $C[0,1]$. Then for i.i.d. sequences by Theorem 1.1.1

$$h(S_n(t))\xrightarrow{\ \mathcal{D}\ }h(W),\quad n\to\infty\ .$$

This explains the term "invariance principle".

1.4 Stationary Random Processes and Fields on the Lattice

The random field $X=\{X_t,\ t\in\mathbb{Z}^\nu\}$, $\nu\ge1$ will be called a translation invariant or stationary (homogenous) one if for arbitrary $k\in\mathbb{N}$, $a,t_1,\dots,t_k\in\mathbb{Z}^\nu$

$$P_X\left(\pi^{-1}_{t_1+a,\dots t_k+a}(B)\right)=P_X\left(\pi^{-1}_{t_1,\dots,t_k}(B)\right),\ B\in\mathcal{B}(S).$$

The random field X_t, $t\in\mathbb{Z}^\nu$ will be called stationary in a wide sense if $EX_t=m$, $E|X_t|^2=b<\infty$, $t\in\mathbb{Z}^\nu$ and for all $a\in\mathbb{Z}^\nu$ the covariance

$$E(X_t-m)(X_{t+a}-m)=R(a)$$

does not depend on $t\in\mathbb{Z}^\nu$.

The function $R(a)$, $a\in\mathbb{Z}^\nu$ is called a correlation function of the random field. It is positive definite. Hence

$$R(a)=\int_{[-\pi,\pi]^k}e^{i(\lambda,a)}dF(\lambda),\ a\in\mathbb{Z}^\nu.$$

The function $F(\lambda)$ is called a spectral function or a spectral measure of the random field. If the spectral measure is absolutely continuous that is

$$F(A)=\int_A f(\lambda)d\lambda,\quad A\in\mathcal{B}([-\pi,\pi]^k),$$

then the function $f(\lambda)$ is called a spectral density.
The spectral function of the stationary in a wide sense sequence X_t, $t\in\mathbb{Z}^1$ is absolutely continuous iff

$$X_t=\sum_{k=-\infty}^{\infty}c_k\,\xi_{k+t}\ ,\qquad\sum_{k=-\infty}^{\infty}|c_k|^2<\infty\ ,\qquad(1.4.1)$$

where the variables ξ_k, $k \in \mathbb{Z}^1$ are orthogonal and $E|\xi_0|^2 = 1$. In this case the spectral density of the sequence X_t, $t \in \mathbb{Z}^1$ has the form

$$f(\lambda) = \frac{1}{2\pi} \left| \sum_{k=-\infty}^{\infty} c_k e^{i\lambda k} \right|^2. \qquad (1.4.2)$$

The variance of the sum $S_n = \sum_{k=1}^{n} X_k$ is expressed in terms of the spectral function $F(\lambda)$ in the following way

$$DS_n = \int_{-\pi}^{\pi} \frac{\sin^2 \frac{n\lambda}{2}}{\sin^2 \frac{\lambda}{2}} \, dF(\lambda).$$

If there exists a spectral density, then

$$DS_n = \int_{-\pi}^{\pi} \frac{\sin^2 \frac{n\lambda}{2}}{\sin^2 \frac{\lambda}{2}} \, f(\lambda) \, d\lambda.$$

If in addition it is continuous in zero, then

$$DS_n = 2\pi f(0) \, n \, (1 + 0(1)), \quad n \to \infty. \qquad (1.4.3)$$

It should be noted that if the random variables $\xi_k, k \in \mathbb{Z}^1$ in (1.4.1) are independent then the random process $X_n, n \in \mathbb{Z}^1$ will be stationary.

To every stationary random process $X_n, n \in \mathbb{Z}^1$ one can associate the measure preserving transformation T of the space of elementary events Ω . By Kolmogorov's theorem as Ω it can be taken the family of functions $\Omega = \left\{ \omega(t), t \in \mathbb{Z}^1 \right\}$ on the one dimensional lattice. The transformation T now is defined in the following way

$$T\left((\omega(t), t \in \mathbb{Z}^1) \right) = (\omega_{t+1}, \, t \in \mathbb{Z}^1).$$

From the stationarity of the process X_t, $t \in \mathbb{Z}^1$ it follows that this transformation preserves the measure of events from σ-field $a_{-\infty}^{\infty} = \sigma(X_t, \, t \in \mathbb{Z}^1)$. By means of the transformation T the process X_t, $t \in \mathbb{Z}^1$ can be represented as

$$X_t(\omega) = X_0(T^{-t}(\omega)), \quad t \in \mathbb{Z}^1.$$

Let $L_2(\Omega)$ be the Hilbert space of square integrable random variables. Here $(\xi, \eta) = E\,\xi\,\bar{\eta}$, $\xi, \eta \in L_2(\Omega)$. Denote by $H(a_{-\infty}^{\infty})$ the subspace of the random variables $\xi \in L_2(\Omega)$ which are measurable with respect to $a_{-\infty}^{\infty}$. Note that the transformation T

considered in this space defines here an unitary operator U acting by the formula

$$(U\xi)(\omega) = \xi(T\omega).$$

1.5 Slowly Varying Functions

The positive function $h(t)$, $t \in [0, \infty)$ is called a slowly varying function if for all $x > 0$

$$\lim_{t \to \infty} \frac{h(tx)}{h(t)} = 1.$$

The slowly varying function has the following properties

1. $\displaystyle\lim_{t \to \infty} \frac{h(t+x)}{h(t)} = 1$, $x \in [0, \infty)$

2. $\displaystyle\lim_{t \to \infty} h(t)t^{-\varepsilon} = 0$, $\displaystyle\lim_{t \to \infty} h(t)t^{\varepsilon} = \infty$, $\varepsilon > 0$

3. $\displaystyle\lim_{k \to \infty} \sup_{2^k \leqslant t \leqslant 2^{k+1}} \frac{h(t)}{h(2^k)} = 1$.

4. If $h(t)$ is integrable in any finite interval, then for each fixed $\alpha > 0$ there exist the constants C_1, $C_2 > 0$ such that

$$\frac{\tau^{\alpha} h(\tau t)}{h(t)} < C_1, \quad \tau t > C_2, \quad \tau \in (0, 1].$$

The positive function $h(n)$ of the natural argument is called slowly varying if for all $k \in \mathbb{N}$

$$\lim_{n \to \infty} \frac{h(kn)}{h(n)} = 1.$$

It could be that the slowly varying function of natural argument does not possess the properties $1 - 4$. However, it is possible to indicate the conditions on which the slowly varying function could be extended on the whole real line, moreover the extended function also will be a slowly varying (see /66/). In this case for a slowly varying function of natural argument the properties 1-4 are valid.

Note, that many other properties of slowly varying functions can be found in /139/.

<u>**1.6 Some Probabilistic Inequalities**</u>

The following theorem estimates the nearness of the distribution functions by the nearness of the corresponding characteristic ones.

<u>Theorem 1.6.1.</u> Let A, T and ε be the constants, $F(x)$ a non-decreasing function, $G(x)$ a real function of bounded variation, $f(t)$ and $g(t)$ corresponding Fourier-Stieltjes transforms such that

1. $F(-\infty) = G(-\infty)$, $F(+\infty) = G(+\infty)$

2. $G'(x)$ exists everywhere and $|G(x)| < A$

3. $\int_{-T}^{T} \left| \frac{f(t)-g(t)}{t} \right| dt = \varepsilon$.

Then for any number $k > 1$ there exists a finite number $c(k)$ only depending on k, such that

$$|F(x) - G(x)| \leq k\frac{\varepsilon}{2\pi} + c(k)\frac{A}{T}.$$

Also we will use the following results.

<u>Theorem 1.6.2.</u> Let $X_1, \ldots, X_n$ be the independent random variables, $EX_i = 0$, $i = \overline{1,n}$ and for some δ, $0 < \delta \leq 1$

$$E|X_j|^{2+\delta} < \infty, \quad j = \overline{1,n}. \qquad \text{Then}$$

$$\sup_X |F_n(x) - \Phi(x)| \leq \frac{C}{B_n^{1+\delta/2}} \sum_{j=1}^{n} E|X_j|^{2+\delta}.$$

Here

$$\sigma_j^2 = EX_j^2, \quad j = \overline{1,n}, \quad B_n = \sum_{j=1}^{n} \sigma_j^2,$$

$$F_n(x) = P\left(\sum_{j=1}^{n} X_j < x\right), \quad \Phi(x) = \frac{1}{\sqrt{2\pi}} \int_{-\infty}^{x} e^{-\frac{t^2}{2}} dt.$$

<u>Theorem 1.6.3.</u> Let X, Y be the random variables and

$F(x) = P(X < x)$, $G(x) = P(X + Y < x)$. Then for any x and $\varepsilon > 0$

$$F(x-\varepsilon) - P(|Y| \geq \varepsilon) \leq G(x) \leq F(x+\varepsilon) + P(|Y| \geq \varepsilon). \qquad (1.6.1)$$

<u>Theorem 1.6.4.</u> Let $X_1, \ldots, X_n$ be independent random variables, $EX_j = 0$, $EX_j^2 < \infty$, $j = \overline{1,n}$. Put

$$V_j(x) = P(X_j < x), \qquad B_n^2 = \sum_{j=1}^{n} EX_j^2 ,$$

$$F_n(x) = P\left(B_n^{-1} \sum_{j=1}^{n} X_j < x \right),$$

$$L_n(\varepsilon) = B_n^{-2} \sum_{j=1}^{n} \int_{|x| \geq \varepsilon B_n} x^2 \, dV_j(x) .$$

Then for any $\varepsilon > 0$

$$\sup_{-\infty < x < \infty} |F_n(x) - \Phi(x)| \leq C(\varepsilon + L_n(\varepsilon)), \qquad C < \infty.$$

<u>Theorem 1.6.5.</u> Let $X_1, \ldots, X_n$ be independent random variables, $EX_j = 0$, $E|X_j|^3 < \infty$. Put $\sigma_j^2 = EX_j^2$, $j = \overline{1,n}$

$$B_n = \sum_{j=1}^{n} \sigma_j^2 , \qquad L_n = B_n^{-\frac{3}{2}} \sum_{j=1}^{n} E|X_j|^3 .$$

Then

$$\left| E e^{it B_n^{-\frac{1}{2}} \sum_{j=1}^{n} X_j} - e^{-\frac{t^2}{2}} \right| \leq 16 L_n |t|^3 e^{-\frac{t^2}{2}}$$

for $|t| \leq \dfrac{1}{4 L_n}$.

1.7 Uniformly Integrable Sequences

The sequence of random variables X_n, $n \in \mathbb{N}$ is said to be uniformly integrable if

$$\lim_{C \to \infty} \sup_{n} \int_{|X_n| \geq C} |X_n| \, P(d\omega) = 0 .$$

Note if for some $\varepsilon > 0$

$$\sup_{n} E|X_n|^{1+\varepsilon} < \infty$$

then the sequence X_n, $n \in \mathbb{Z}^1$ is uniformly integrable.

<u>Theorem 1.7.1.</u> Suppose the sequence X_n, $n \in \mathbb{Z}^1$ is uniformly integrable. Then $X_n \xrightarrow{\mathcal{D}} X$ implies $EX_n \to EX$. If X_n $n \in \mathbb{N}$ and X are non-negative and integrable then $X_n \xrightarrow{\mathcal{D}} X$ is equivalent to $EX_n \to EX$, $n \to \infty$.

2. WEAK DEPENDENCE CONDITIONS FOR RANDOM PROCESSES AND FIELDS

2.1 Measures of Dependence

Let $(\Omega, \mathcal{F}, P)$ be a probability space. There exist many various definitions of the measures of dependence between σ - fields $\mathcal{O}t_1$, $\mathcal{O}t_2 \subset \mathcal{F}$. Denote

$$\alpha(A, B) = P(AB) - P(A)P(B), \quad \varphi(A, B) = P(^A\!/_B) - P(A),$$

$$\psi(A, B) = \frac{P(AB)}{P(A)P(B)} - 1, \quad A, B \in \mathcal{F}.$$

We will be **mainly interested in** the following coefficients of dependence (mixing) between σ-fields:

1) strong mixing (s.m.) or α-mixing coefficient

$$\alpha(\mathcal{O}t_1, \mathcal{O}t_2) = \sup \{ |\alpha(A, B)|, \ A \in \mathcal{O}t_1, \ B \in \mathcal{O}t_2 \},$$

2) uniformly strong mixing (u.s.m.) or φ-mixing coefficient

$$\varphi(\mathcal{O}t_1, \mathcal{O}t_2) = \sup \{ |\varphi(A, B)|, \ A \in \mathcal{O}t_1, B \in \mathcal{O}t_2, P(B) > 0 \},$$

3) ψ-mixing coefficient

$$\psi(\mathcal{O}t_1, \mathcal{O}t_2) = \sup \{ |\psi(A, B)|, \ A \in \mathcal{O}t_1, B \in \mathcal{O}t_2, P(A), P(B) > 0 \},$$

4) coefficient of complete regularity or ρ-mixing coefficient

$$\rho(\mathcal{O}t_1, \mathcal{O}t_2) = \sup \{ |EXY - EXEY| \left(\sqrt{E(X - EX)^2 E(Y - EY)^2} \right)^{-1},$$
$$X \in L_2(\mathcal{O}t_1), \ Y \in L_2(\mathcal{O}t_2) \}.$$

Note that each of the introduced coefficients assumes the value o when the σ-fields $\mathcal{O}t_1$ and $\mathcal{O}t_2$ are independent. There is the following hierarchy between these mixing coefficients

$$2 \alpha(\mathcal{O}t_1, \mathcal{O}t_2) \leqslant \varphi(\mathcal{O}t_1, \mathcal{O}t_2) \leqslant \tfrac{1}{2} \psi(\mathcal{O}t_1, \mathcal{O}t_2) \qquad (2.1.1)$$

$$2\alpha(\mathcal{O}_1, \mathcal{O}_2) \leq \rho(\mathcal{O}_1, \mathcal{O}_2) \leq \psi(\mathcal{O}_1, \mathcal{O}_2), \qquad (2.1.2)$$

$$\rho(\mathcal{O}_1, \mathcal{O}_2) \leq 2\varphi^{1/2}(\mathcal{O}_1, \mathcal{O}_2). \qquad (2.1.3)$$

Also the following inequalities are valid:

$$\alpha(\mathcal{O}_1, \mathcal{O}_2) \leq \tfrac{1}{4}, \quad \varphi(\mathcal{O}_1, \mathcal{O}_2) \leq 1, \quad \rho(\mathcal{O}_1, \mathcal{O}_2) \leq 1. \qquad (2.1.4)$$

Before **proving** these relations we give some important estimates for

$$Cov(X, Y) = EXY - EXEY,$$

the covariance of random variables X, Y assumed measurable with respect to the corresponding σ-fields.

<u>Lemma 2.1.1.</u> (/66/) Let X, Y be random variables measurable with respect to σ-fields $\mathcal{O}_1$ and $\mathcal{O}_2$ respectively. If

$$E|X|^{\tau}, \ E|Y|^{s} < \infty, \quad \tau, s \geq 1, \quad \tfrac{1}{\tau} + \tfrac{1}{s} + \varepsilon = 1, \quad \varepsilon \geq 0,$$

then

1) $|Cov(X,Y)| \leq C E^{\frac{1}{\tau}}|X|^{\tau} E^{\frac{1}{s}}|Y|^{s} \alpha^{\varepsilon}(\mathcal{O}_1, \mathcal{O}_2), \quad 0 < C < \infty,$

2) $|Cov(X,Y)| \leq 2 E^{\frac{1}{\tau}}|X|^{\tau} E^{\frac{1}{s}}|Y|^{s} \varphi^{\frac{1}{\tau} + \varepsilon}(\mathcal{O}_1, \mathcal{O}_2)$
and

3) $|Cov(X,Y)| \leq E|X| E|Y| \psi(\mathcal{O}_1, \mathcal{O}_2).$

The proof of 1) will be obtained in Chapter 6 as a special case of the more general estimate. The estimate 3) is evident in the case of simple random variables. In general case 3) follows from the possibility of approximation the random variables X and Y by simple ones. Now we give the proof of estimate 2).

<u>Proof.</u> First suppose that the random variables X and Y are simple, i.e.,

$$X = \sum_{i=1}^{N} \lambda_i I_{A_i}, \quad A_i \in \mathcal{O}_1, \quad A_i \cap A_m = \varnothing, \quad \bigcup_{i=1}^{N} A_i = \Omega,$$

$$Y = \sum_{j=1}^{M} \mu_j I_{B_j}, \quad B_j \in \mathcal{O}_2, \quad B_j \cap B_k = \varnothing \quad \bigcup_{j=1}^{M} B_j = \Omega, \qquad (2.1.5)$$

where $I_{(\cdot)}$ is the indicator of a set, $P(A_i), P(B_j) > 0$,

$$i = \overline{1, N}, \quad j = \overline{1, M}.$$

Then by the Hölder inequality we have

$$| \text{Cov}(X,Y)| \leq \sum_{i=1}^{N} \sum_{j=1}^{M} |\lambda_i||\mu_j| \alpha(A_i, B_j) =$$

$$= \sum_{i=1}^{N} |\lambda_i| P^{1/\tau}(A_i) P^{1/s+\varepsilon}(A_i) \sum_{j=1}^{M} |\mu_j||\varphi(B_j, A_i)| \leq \qquad (2.1.6)$$

$$\leq \left(\sum_{i=1}^{N} |\lambda_i|^{\tau} P(A_i) \right)^{1/\tau} \left(\sum_{i=1}^{N} P(A_i) \left(\sum_{j=1}^{M} |\mu_j||\varphi(B_j, A_i)| \right)^{\frac{s}{1+s\varepsilon}} \right)^{\frac{1+s\varepsilon}{s}}$$

and

$$\sum_{j=1}^{M} |\mu_j||\varphi(B_j, A_i)| = \sum_{j=1}^{M} |\mu_j||\varphi(B_j, A_i)|^{1/s} |\varphi(B_j, A_i)|^{1/\tau+\varepsilon} \leq$$

$$\leq \left(\sum_{j=1}^{M} |\mu_j|^{s} |\varphi(B_j, A_i)| \right)^{1/s} \left(\sum_{j=1}^{M} |\varphi(B_j, A_i)| \right)^{\frac{1+\tau\varepsilon}{\tau}} .$$

Hence

$$\sum_{i=1}^{N} P(A_i) \left(\sum_{j=1}^{M} |\mu_j||\varphi(B_j, A_i)| \right)^{\frac{s}{1+s\varepsilon}} \leq$$

$$\sum_{i=1}^{N} P(A_i) \left(\sum_{j=1}^{M} |\mu_j|^{s} |\varphi(B_j, A_i)| \right)^{\frac{1}{1+s\varepsilon}} \left(\sum_{j=1}^{M} |\varphi(B_j, A_i)| \right)^{\frac{(1+\tau\varepsilon)s}{(1+s\varepsilon)\tau}} =$$

$$\sum_{i=1}^{N} P^{\frac{1}{1+s\varepsilon}}(A_i) \left(\sum_{j=1}^{M} |\mu_j|^{s} |\varphi(B_j, A_i)| \right)^{1/1+s\varepsilon} (P(A_i))^{\frac{s\varepsilon}{1+s\varepsilon}} \times$$

$$\times \left(\sum_{j=1}^{M} |\varphi(B_j, A_i)| \right)^{\frac{s(1+\tau\varepsilon)}{\tau(1+s\varepsilon)}} \leq$$

$$\leq \left(\sum_{i=1}^{N} P(A_i) \right) \left(\sum_{j=1}^{M} |\mu_j|^{s} |\varphi(B_j, A_i)| \right)^{1/1+s\varepsilon} \times$$

$$\times \left(\sum_{i=1}^{N} P(A_i) \left(\sum_{j=1}^{M} |\varphi(B_j, A_i)| \right)^{\frac{1+\tau\varepsilon}{\tau\varepsilon}} \right)^{\frac{s\varepsilon}{1+s\varepsilon}} . \qquad (2.1.7)$$

Substituting (2.1.6) in (2.1.5) we obtain

$$|\operatorname{Cov}(X,Y)| \leq E^{1/r}|X|^{r}\left(\sum_{i=1}^{N}P(A_i)\left(\sum_{j=1}^{M}|\mu_j|^{s}|\varphi(B_j,A_i)|\right)\right)^{1/s} \times$$

$$\times \left(\sum_{i=1}^{N}P(A_i)\left(\sum_{j=1}^{M}|\varphi(B_j,A_i)|\right)^{\frac{1+r\varepsilon}{r\varepsilon}}\right)^{\varepsilon}.$$

Note that

$$\sum_{i=1}^{N}P(A_i)\left(\sum_{j=1}^{M}|\mu_j|^{s}|\varphi(B_j,A_i)|\right) \leq 2E|Y|^{s}$$

and

$$\sum_{j=1}^{M}|\varphi(B_j,A_i)| \leq 2\varphi(\mathcal{O}_1,\mathcal{O}_2).$$

Then

$$|\operatorname{Cov}(X,Y)| \leq 2E^{1/r}|X|^{r}E^{1/s}|Y|^{s}\varphi^{1/r+\varepsilon}(\mathcal{O}_1,\mathcal{O}_2). \qquad (2.1.8)$$

Since one can always construct sequences of simple random variables $\{X_n\}$, $\{Y_n\}$ of the type (2.1.5) such that

$$E|X_n-X|^{r}, \quad E|Y_n-Y|^{s} \to 0, \quad n \to \infty$$

then the estimate 2) of Lemma 2.1.1 follows from (2.1.8).

<u>Corollary 2.1.1.</u> Let X, Y be random variables measurable with respect to $\mathcal{O}_1$, $\mathcal{O}_2$ respectively, and with probability 1, $|Y| \leq C_1 < \infty$.
Then

$$|\operatorname{Cov}(X,Y)| \leq 2C_1 E|X|\varphi(\mathcal{O}_1,\mathcal{O}_2),$$

$$|\operatorname{Cov}(X,Y)| \leq B_1 E^{1/r}|X|^{r}\alpha^{1-1/r}(\mathcal{O}_1,\mathcal{O}_2), \quad 0 < B_1 < \infty.$$

If in addition with probability 1, $|X| \leq C_2 < \infty$ then

$$|\operatorname{Cov}(X,Y)| \leq 2C_1 C_2 \varphi(\mathcal{O}_1,\mathcal{O}_2),$$

$$|\operatorname{Cov}(X,Y)| \leq B_2 \alpha(\mathcal{O}_1,\mathcal{O}_2), \quad 0 < B_2 < \infty.$$

Indeed, putting $r = \infty$, $s = 1$ or $r = \infty$, $s = \infty$, $\varepsilon = 1$ in the estimates 1) and 2) of Lemma 2.1.1 we obtain the corollary by formal way. However, one can establish this result directly by looking over the corresponding proofs of Lemma 2.1.1. Now prove the relations (2.1.1)-(2.1.4). For any $A \in \mathcal{O}_1$,

$B \in \mathcal{O}_2$, $0 < P(A), P(B) < 1$ we can write

$$|\alpha(A,B)| = |\alpha(A,\bar{B})| \leq \min\left\{|\varphi(A,B)|P(B),\ |\varphi(A,\bar{B})|P(\bar{B})\right\} \leq$$

$$\leq \varphi(\mathcal{O}_1, \mathcal{O}_2)\,\min\left\{P(B), P(\bar{B})\right\} \leq \tfrac{1}{2}\,\varphi(\mathcal{O}_1, \mathcal{O}_2).$$

Hence

$$\alpha(\mathcal{O}_1, \mathcal{O}_2) \leq \tfrac{1}{2}\,\varphi(\mathcal{O}_1, \mathcal{O}_2).$$

Further

$$|\varphi(A,B)| = |\varphi(\bar{A},B)| \leq \min\left\{|\psi(A,B)|P(A),\ |\psi(\bar{A},B)|P(\bar{A})|\right\} \leq$$

$$\leq \psi(\mathcal{O}_1, \mathcal{O}_2)\,\min\left\{P(A), P(\bar{A})\right\} \leq \tfrac{1}{2}\,\psi(\mathcal{O}_1, \mathcal{O}_2).$$

Hence

$$\varphi(\mathcal{O}_1, \mathcal{O}_2) \leq \tfrac{1}{2}\,\psi(\mathcal{O}_1, \mathcal{O}_2)$$

and the relations (2.1.1) are proven. Put $\xi = I_A, \eta = I_B, A \in \mathcal{O}_1, B \in \mathcal{O}_2$.
It is evident that $\xi \in L_2(\mathcal{O}_1)$, $\eta \in L_2(\mathcal{O}_2)$.
Then

$$|\alpha(A,B)| = |\,\mathrm{Cov}(\xi,\eta)| \leq \rho(\mathcal{O}_1, \mathcal{O}_2)\sqrt{D\xi\,D\eta} =$$

$$= \rho(\mathcal{O}_1, \mathcal{O}_2)\sqrt{P(A)P(\bar{A})P(B)P(\bar{B})} \leq \tfrac{1}{4}\,\rho(\mathcal{O}_1, \mathcal{O}_2).$$

Thus

$$\alpha(\mathcal{O}_1, \mathcal{O}_2) \leq \tfrac{1}{2}\,\rho(\mathcal{O}_1, \mathcal{O}_2).$$

To prove (2.1.2) it remains to note that the inequality
$\rho(\mathcal{O}_1, \mathcal{O}_2) \leq \psi(\mathcal{O}_1, \mathcal{O}_2)$ follows from Lemma 2.1.1 and Lyapu-
nov's inequality. The relation (2.1.3) follows directly from
Lemma 2.1.1. The first estimate in (2.1.4) is the consequence
of the following inequality

$$\alpha(A,B) \leq \sqrt{P(A)P(\bar{A})P(B)P(\bar{B})} \leq \tfrac{1}{4}.$$

The other estimates in (2.1.4) are evident.

2.2 Weak Dependence Conditions for Random Processes.
Classical Examples

Let ξ_t , $t \in \mathbb{Z}^1$ be a stationary random process (s.r.p.) and
$\mathcal{O}_T = \mathfrak{S}(\xi_t,\ t \in T)$, $T \subset \mathbb{Z}^1$ be the $\mathfrak{S}$-field generated
by corresponding random variables. Historically the first weak

dependence condition for the s.r.p. which was successfully
used for the proof of c.l.th. was Rosenblatt's s.m. (or α -
mixing) condition (see /111/).

It is said that the s.r.p. ξ_t, $t \in \mathbb{Z}^1$, satisfies the α -mixing
condition if

$$\alpha\left(\mathcal{O}_{]-\infty,0]},\ \mathcal{O}_{[n,+\infty)}\right) = \alpha(n) \to 0, \quad n \to \infty, \quad n \in \mathbb{N}.$$

Later there were also introduced other weak dependence conditions, namely φ -mixing or u.s.m. condition of Ibragimov (/63/):

$$\varphi\left(\mathcal{O}_{]-\infty,0]},\ \mathcal{O}_{[n,+\infty)}\right) = \varphi(n) \to 0, \quad n \to \infty, \quad n \in \mathbb{N},$$

ψ -mixing condition (/8/) ,

$$\psi\left(\mathcal{O}_{]-\infty,0]},\ \mathcal{O}_{[n,+\infty)}\right) = \psi(n) \to 0, \quad n \to \infty, \quad n \in \mathbb{N}$$

and ρ -mixing condition(/42/),

$$\rho\left(\mathcal{O}_{]-\infty,0]},\ \mathcal{O}_{[n,+\infty)}\right) = \rho(n) \to 0, \quad n \to \infty, \quad n \in \mathbb{N}.$$

For singly-infinite processes ξ_t, $t \in \mathbb{N} \cup \{0\}$ in the above
mentioned mixing conditions one can use the σ -fields
$\mathcal{O}_{[0,\kappa]}$ and $\mathcal{O}_{[n+\kappa,+\infty)}$, $\kappa,\ n \in \mathbb{N}.$

Denote by $\Pi_{j,f}^{(\kappa)}$, $k = 1,2$, the class of s.r.p. satisfying
the j -mixing condition, $j \in (\alpha, \varphi, \psi, \rho)$, $|j(n)| \leqslant f(n)$,
$n \in \mathbb{N}$, and the components of which has m -th order moments
($0 < m \leqslant \kappa$, $m \in R_+^1$). The following inclusions directly come
from the inequalities (2.1.1) − (2.1.3)

$$\Pi_{\psi,f}^{(\kappa)} \subset \Pi_{\varphi,\,\frac{1}{2}f}^{(\kappa)} \subset \Pi_{\alpha,\,\frac{1}{4}f}^{(\kappa)} \ ,$$

$$\Pi_{\psi,f}^{(\kappa)} \subset \Pi_{\rho,f}^{(\kappa)} \subset \Pi_{\alpha,\,\frac{1}{2}f}^{(\kappa)} \ ,$$

$$\Pi_{\varphi,f^2}^{(\kappa)} \subset \Pi_{\rho,2f}^{(\kappa)} \ , \qquad \kappa = 0, 1, 2 \ldots$$

The examples of s.r.p. which show the strong inclusion in these
relations will be given later.

One of the important questions about various mixing conditions
is how broad is the class of processes which satisfy these con-

ditions. There are several well known examples of the random
processes with weakly dependent components. Some of them we
describe below (the analysis of these examples can be found
for instance in   /7/   or   /66/). As a basic class we
will consider the Gibbs random fields. Their theory will be
presented in Chapter 9.

1. Sequence of m -dependent Random Variables

The sequence ξ_t, $t \in \mathbb{Z}^1$ is called m -dependent if the
σ -fields $\sigma(\xi_{t+j}, j=1,...,a)$ and $\sigma(\xi_{s-i}, i=1,...,\beta)$
are independent if $t - s > m$, $m \in \mathbb{N}$. For instance if $f(x_1,...,x_m)$
is some Borel function of m -variables and η_t, $t \in \mathbb{Z}^1$
is a family of independent random variables, then the sequence
$\xi_t = f(\eta_t,...,\eta_{t+m})$ is m -dependent. It is obvious that
a sequence of m -dependent random variables is j -mixing,
$j \in (\alpha, \varphi, \psi, \rho)$. A profound study of such sequences is given
in /138/.

2. Markov Chains

a) Let ξ_t, $t \in \mathbb{Z}^1$ be a Markov chain which takes values in
some space X . Such chain is defined by the probability
transition function $p(x, A)$, $x \in X$, $A \in \mathcal{F}_X$ and the initial distribution $\pi(A)$, $A \in \mathcal{F}_X$, where $\mathcal{F}_X$ is the σ -
field of subsets of X . Suppose that there exists a finite
measure m on $\mathcal{F}_X$, $m(X) > 0$, and the numbers $k \in \mathbb{N}$,
$\varepsilon \in R^1$, $k > 1$, $\varepsilon > 0$ such that if $m(A) \leq \varepsilon$ then
$p^{(k)}(x, A) \leq 1 - \varepsilon$ (Doeblin's condition). Suppose also that
the considered Markov chain has only one ergodic class without
subclasses. Then (/36/) there exists a distribution $p(A)$,
$A \in \mathcal{F}_X$ such that

$$\sup_{x, A} | p^{(n)}(x, A) - p(A) | \leq C \tau^n, \quad C > 0, \quad 0 < \tau < 1.$$

Now it is not difficult to prove that under $p(A)$, $A \in \mathcal{F}_X$,
as initial distribution the sequence ξ_t, $t \in \mathbb{Z}^1$ is stationary
and φ -mixing. Moreover

$$\varphi(n) \leq C_1 \tau^n, \quad C_1 > 0, \quad 0 < \tau < 1.$$

b) Let ξ_t, $t \in \mathbb{Z}^1$ be a homogeneous Markov chain with count-
able state space X and transition matrix $(p_{ij})i, j \in X$.
Suppose that the considered Markov chain has a single aperiodic
positive recurrent class. Then there exists the limit

$$\lim_{n \to \infty} P_{ij}^{(n)} = \pi_{ij} , \quad i, j \in X$$

24

and π_j , $j \in X$ is the unique stationary distribution of this chain. Moreover, if π_j , $j \in X$ is taken as the initial distribution then the r.p. ξ_t , $t \in \mathbb{Z}^1$, is stationary and α -mixing with

$$\alpha(n) \leq C \sum_{i \in X} \pi_i \sum_{j \in X} | p_{ij}^{(n)} - \pi_j |, \qquad 0 < C < \infty.$$

c) Let ξ_t , $t \in \mathbb{Z}^1$ be a stationary Markov chain with finite state space and irreducible aperiodic transition matrix. Then

$$\psi(n) \leq B\tau^n, \qquad B > 0, \quad 0 < \tau < 1.$$

3. Levy's example

Let $\Omega = [0,1]$ be the unit interval with the Gaussian measure

$$P(A) = \frac{1}{\lg 2} \int_A \frac{dx}{1+x} , \quad A \in \mathcal{B}$$

on the $\mathfrak{S}$ -field $\mathcal{B}$ of its Borel subsets. Each real number $t \in \Omega$ has the unique continued-fraction expansion

$$t = \cfrac{1}{a_1(t) + \cfrac{1}{a_2(t) + \dots}} .$$

The sequence of random variables $a_1(t), a_2(t) \dots$ is stationary and φ -mixing with

$$\varphi(n) \leq C\tau^n, \quad C > 0, \quad 0 < \tau < 1 .$$

2.3 Davydov's Examples

Davydov's examples /25/ are constructed as some functionals of the specific Markov chain first considered by Chung /21/ . The Chung construction represents the Markov chain with state $\mathbb{Z}^1$ and transition matrix such that

$$p_{n,n+1} = p_{-n,-n-1} = a_n , \ n \geq 0; \quad p_{n,0} = p_{-n,0} = 1 - a_n ,$$

$$n > 0; \quad p_{00} = 0, \quad a_0 = \frac{1}{2} , \quad 0 < a_n < 1, \quad n \geq 1.$$

Suppose $f_{00}^{(n)}$ denotes the probability of the following event: the chain after leaving the origin returns there for the first time on the n -th step. Then

$$f_{00}^{(n)} = \beta_{n-1} - \beta_n , \qquad \beta_0 = \beta_1 = 1 , \qquad \beta_n = \prod_{j=1}^{n} a_j.$$

This implies that the class will be recurrent iff $\beta_n \to 0$. The stationary distribution exists iff $\sum_{n=0}^{\infty} \beta_n < \infty$ and has a form

$$\pi_0 = \left(\sum_{n=0}^{\infty} \beta_n \right)^{-1}, \quad \pi_j = \pi_{-j} = \tfrac{1}{2} \pi_0 \beta_{j-1}, \quad j > 0.$$

Note also the following evident statement.

Proposition 2.3.1. Let the s.r.p. ξ_t, $t \in \mathbb{Z}^1$, satisfy the j-mixing condition, $j \in \{\alpha, \varphi, \psi, \rho\}$ and $g(x)$, $x \in R^1$, be some Borel-measurable function. Then the s.r.p. $\eta_t = g(\xi_t)$, $t \in \mathbb{Z}^1$ is j-mixing at least with the same mixing coefficient.

Example 1. Put $g(k) = \operatorname{sing}\{k\}|k|^{1/\tau}$, $k \in \mathbb{Z}^1$, $\tau \in R^1_+$,

$$f_{00}^{(n)} = a_1 \dots a_{n-2}(1 - a_{n-1}) = C n^{-2-\delta}, \quad \delta > 1, \quad \delta \tau > 2,$$

where the constant C is chosen from condition

$$\sum_{n=1}^{\infty} f_{00}^{(n)} = 1.$$

Taking as the initial distribution a stationary one and putting $\eta_t = g(\xi_t)$, $t \in \mathbb{Z}^1$, we obtain s.r.p. satisfying the α-mixing condition. This example as well as the next one is interesting by the non-standard behaviour of the variance of the sum $S_n = \sum_{t=1}^{n} \eta_t$ namely,

$$DS_n \asymp n^{2-\delta+\frac{2}{\tau}}, \quad n \in \mathbb{N}. \tag{2.3.1}$$

It is said that the variance DS_n has the standard or regular behaviour if $DS_n = n\,h(n)$, where $h(n)$ is a slowly varying function. Davydov also calculates the exact rate of decay of the mixing coefficient $\alpha(n)$ in this example:

$$\alpha(n) \asymp n^{-\delta}.$$

The proof of this fact (to be found in /25/) is long and will not be presented here. Let us prove the relation (2.3.1). Since the Markov chain used in Davydov's example is symmetric and the function $g(k)$ is odd then we can write

$$R(n) = \sum_{k,s \in \mathbb{Z}^1} f(k) f(s) \pi_k p_{k,s}(n) =$$

$$= 2 \sum_{k,s \in \mathbb{N}} f(k) f(s) \pi_k \left(p_{k,s}(n) - p_{k,-s}(n) \right).$$

It is not difficult to check that

$$P_{k,s}(n) - P_{k,-s}(n) = \begin{cases} a_k\, a_{k+1} \cdots a_{k+n-1}, & s = k+n, \\ 0, & s \neq k+n. \end{cases}$$

Hence

$$R(n) = 2 \sum_{k \in \mathbb{N}} g(k)\, g(k+n)\, \pi_{k+n}.$$

Since $\quad \pi_k \asymp k^{-(1+\delta)} \quad$, $\quad g(k) = k^{1/2} \quad$, $k \in \mathbb{N}$, then

$$R(n) \asymp 2 \sum_{k \in \mathbb{N}} k^{1/2} (k+j)^{-(1+\delta-\frac{1}{2})} \asymp j^{\frac{2}{2}-\delta}.$$

Therefore

$$DS_n = n D \xi_0 + 2 \sum_{j=1}^{n-1} (n-j) R(j) \asymp n^{2-\delta+\frac{2}{2}}.$$

<u>Example 2.</u> The probability $\quad f_{00}^{(n)} \quad$ is defined as in Example 1,
$0 < \delta < 1$,

$$g(0) = 0, \quad g(k) = -g(-k) = 1 + k^{-1}, \quad k \in \mathbb{N}.$$

It follows by the same argument as was used in Example 1 that

$$DS_n \asymp n^{2-\delta}.$$

2.4 Herrndorf's Examples

The class of examples constructed by Herrndorf has several
interesting properties (see /58/).
For any sequence of the real numbers ε_n , $n \in \mathbb{N}$, by Herrndorf's procedure one can construct the s.r.p. ξ_n , $n \in \mathbb{N}$, such that

$$\alpha(n) \leqslant \varepsilon_n, \quad E \xi_1 = 0, \quad 0 < E \xi_1^2 < \infty, \quad E \xi_1 \xi_n = 0, \quad \text{and}$$

the partial sums $\quad S_n = \sum_{t=1}^{n} \xi_t \quad$ have the following properties:

1) $\displaystyle \inf_{n \in \mathbb{N}} P(S_n = 0) > 0$

2) the family of partial sums S_n , $n \in \mathbb{N}$, is tight

3) $S_n\, b_n^{-1} \to 0$ in probability as $n \to \infty$

for any sequence b_n , $n \in \mathbb{N}$, such that $b_n \in (0, +\infty)$ and
$b_n \to \infty$, $n \to \infty$.
The importance of these properties of Herrndorf's examples for
the analysis of certain aspects of c.l.th. for s.r.p. will be
clear from Chapter 5.

Before constructing the examples we prove some preliminary Lemmas, where we mostly follow /58/ .

<u>Lemma 2.4.1.</u> Let F be an atom of the σ - field .Then

$$\alpha\,(\mathcal{F},\,\mathcal{F}\,)\leqslant 1-P(F).$$

<u>Proof.</u> Let $A,B\in\mathcal{F}$. If $A\cap F=\emptyset$ or $B\cap F=\emptyset$ then

$$0\leqslant P(AB)\leqslant 1-P(F),$$
$$0\leqslant P(A)P(B)\leqslant 1-P(F).$$

If $A\supset F$, $B\supset F$ then

$$P(F)-1\leqslant\alpha(A,B)\leqslant P(A)(1-P(B))\leqslant 1-P(F).$$

Hence

$$|\alpha\,(A,B)|\leqslant 1-P(F).$$

<u>Lemma 2.4.2</u> (/12/). Let $\mathcal{F}_n$, $\mathcal{B}_n$, $n\in\mathbb{N}$, be σ-fields. Then

$$\alpha(\bigvee_{n=1}^{\infty}\mathcal{F}_n,\;\bigvee_{n=1}^{\infty}\mathcal{B}_n)\leqslant\sum_{n=1}^{\infty}\sum_{m=1}^{\infty}\alpha(\mathcal{F}_n,\,\mathcal{B}_m).$$

<u>Proof.</u>

First we show that

$$\alpha(\mathcal{F}_1\vee\mathcal{F}_2,\,\mathcal{B}_1)\leqslant\alpha(\mathcal{F}_1,\,\mathcal{B}_1)+\alpha(\mathcal{F}_2,\,\mathcal{B}_2). \qquad (2.4.1)$$

Let $B\in\mathcal{B}_1$ be fixed. Then for arbitrary $A_1\in\mathcal{F}_1$, $A_2\in\mathcal{F}_2$ we have

$$|\alpha(A_1,B)|\leqslant\alpha(\mathcal{F}_1,\mathcal{B}_1),\qquad |\alpha(A_2,B)|\leqslant\alpha(\mathcal{F}_2,\mathcal{B}_1).$$

Thus if $A\in\mathcal{F}_1\cup\mathcal{F}_2$, then

$$|\alpha(A,B)|\leqslant\alpha(\mathcal{F}_1,\mathcal{B}_1)+\alpha(\mathcal{F}_2,\mathcal{B}_2).$$

Hence for any $A\in\mathcal{F}_1\vee\mathcal{F}_2$, $B\in\mathcal{B}_1$

$$|\alpha(A,B)|\leqslant\alpha(\mathcal{F}_1,\mathcal{B}_1)+\alpha(\mathcal{F}_2,\mathcal{B}_2),$$

i.e.,(2.4.1) is valid. According to the symmetry

$$\alpha(\mathcal{F}_1\vee\mathcal{F}_2,\,\mathcal{B}_1\vee\mathcal{B}_2)\leqslant\alpha(\mathcal{F}_1,\mathcal{B}_1)+\alpha(\mathcal{F}_1,\mathcal{B}_2)+\alpha(\mathcal{F}_2,\mathcal{B}_1)+\alpha(\mathcal{F}_2,\mathcal{B}_2).$$

To complete the proof it remains to use the induction by n .

<u>Lemma 2.4.3.</u> Let ξ_n , $n\in\mathbb{Z}^1$, be the i.i.d. random variables with $P(\xi_n=0)\geqslant 1-\varepsilon$ for some $0\leqslant\varepsilon\leqslant 1$. Let

$$X_n=\sum_{j=-N}^{N}c_j\,\xi_{j+n}\,,\quad n\in\mathbb{N}\cup\{0\}\,,\quad c_j\in R^1.$$

Then for the process X_n , $n \in \mathbb{N}$,

$$\alpha(n) \leq \varepsilon \max(2N - n + 1, 0).$$

<u>Proof.</u>   By using the previous lemma we have

$$\alpha(n) \leq \alpha(\sigma(\xi_i, i \leq N), \sigma(\xi_i, i \geq n - N)) \leq \sum_{i=n-N}^{N} \alpha(\sigma(\xi_i), \sigma(\xi_i)).$$

Since the event $\{\xi_n = 0\}$ is an atom, then by Lemma 2.4.1

$$\alpha(\sigma(\xi_i), \sigma(\xi_i)) \leq \varepsilon.$$

This implies the validity of Lemma 2.4.3.

Lemma 2.4.3 shows that even if $N = N(n) \to \infty$, as $n \to \infty$, then one can get $\alpha(n) \to 0$, as $n \to \infty$ by taking an increasing sequence of probabilities $P(\xi_n = 0)$, $n \in \mathbb{N}$. Herrndorf's examples have the following structure:

$$X_n = \sum_{k \in \mathbb{N}} X_{n,k} , \quad n \in \mathbb{N} , \qquad (2.4.2)$$

$$X_{n,k} = \sum_{j=-N(k)}^{N(k)} c_{jk} \, \xi_{j+n,k} , \qquad (2.4.3)$$

where ξ_{jk} , $j \in \mathbb{Z}^1$, $k \in \mathbb{N}$, is the family of independent random variables such that $E\xi_{jk} = 0$, $D\xi_{jk} = 1$ and the distribution of which does not depend on j , $j \in \mathbb{N}$ and has a "heavy point" in the origin. It is clear that for any fixed n the sequence $X_{n,k}$, $k \in \mathbb{N}$ has independent components. The important technical lemma stated below permits to choose the coefficients c_{jk} and the distribution of random variables $j \in \mathbb{Z}^1$, $k \in \mathbb{N}$ such that the constructions (2.4.2), (2.4.3) make sense and the above formulated statements 1)-3) are fulfilled.

<u>Lemma 2.4.4.</u> There exist functions $f_k(\lambda)$, $\lambda \in [-\frac{1}{2}, \frac{1}{2}]$, such that

$$f_k(\lambda) = \left| \sum_{j=-N(k)}^{N(k)} c_{jk} \, e^{2\pi i j \lambda} \right|^2, \qquad N(k) \in \mathbb{N},$$

$$c_{jk} \in R^1, \qquad c_{-j,k} = c_{j,k} , \qquad \sum_{j=-N(k)}^{N(k)} c_{jk} = 0$$

and

$$\sum_{k \in \mathbb{N}} f_k(\lambda) = 1 , \qquad \lambda \in [-\frac{1}{2}, \frac{1}{2}].$$

<u>Proof.</u> The sequence f_k, $k \in \mathbb{N}$ will be defined by induction. Put $f_1 = 0$. Suppose that $f_1, \ldots, f_n$ have already been const-

ructed and have the following properties

$$\sum_{k=1}^{n} f_k(\lambda) \le 1 - 2^{-n}, \qquad \lambda \in [-\tfrac{1}{2}, \tfrac{1}{2}], \qquad\qquad (2.4.4)$$

$$\sum_{k=1}^{n} f_k(\lambda) \ge 1 - 2^{-n+1}, \qquad \lambda \in [-\tfrac{1}{2}, \tfrac{1}{2}] \setminus [-2^{-n}, 2^{-n}]. \qquad (2.4.5)$$

Let

$$h(\lambda) = \min(1, |\lambda| 2^{n+1}),$$

$$g(\lambda) = \left(1 - 2^{-n} + 2^{-n-2} - \sum_{k=1}^{n} f_k(\lambda)\right) h(\lambda).$$

By virtue of (2.4.4) $g(\lambda) \ge 0$. It is evident from the properties of the function $g(\lambda)$ and $f_k(\lambda)$, $k = \overline{1,n}$ that $\sqrt{g(\lambda)}$ is a continuous even function. Hence it can be uniformly approximated on $[-\tfrac{1}{2}, \tfrac{1}{2}]$ by trigonometrical polynomials. Thus for any $\varepsilon > 0$ there exists $N \in \mathbb{N} \cup \{0\}$ and $a_j \in R^1$ such that

$$\left| \sqrt{g(\lambda)} - \sum_{j=0}^{N} a_j \cos(2\pi j \lambda) \right| \le \varepsilon, \qquad \lambda \in [-\tfrac{1}{2}, \tfrac{1}{2}]. \qquad (2.4.6)$$

In particular, since $g(0) = 0$, then $\left| \sum_{j=0}^{N} a_j \right| \le \varepsilon$.

Put

$$\varepsilon = 2^{-n-4}, \quad N(n+1) = N, \quad c_{j,n+1} = a_{|j|} 2^{-1},$$

$$j \in \{1, \ldots, N\} \cup \{-1, \ldots, -N\}, \quad c_{0,n+1} = a_0 - \sum_{j=0}^{N} a_j$$

and define $f_{n+1}(\lambda)$ with these parameters.

Let us verify that both (2.4.4) and (2.4.5) remain valid when we replace n by $n+1$. Using (2.4.6) we get

$$\left| \sqrt{g(\lambda)} - \sum_{j=-N}^{N} c_{j,n+1} e^{2\pi i j \lambda} \right| =$$

$$= \left| \sqrt{g(\lambda)} - \sum_{j=0}^{N} a_j \cos 2\pi j \lambda + \sum_{j=0}^{N} a_j \right| \le 2\varepsilon.$$

Hence

$$\left| g(\lambda) - f_{n+1}(\lambda) \right| = \left| g(\lambda) - \left(\sum_{j=0}^{N} a_j (1 - \cos 2\pi j \lambda) \right)^2 \right| \le 4\varepsilon. \qquad (2.4.7)$$

Suppose that $|\lambda| 2^{n+1} \ge 1$, $\qquad \lambda \in [-\tfrac{1}{2}, \tfrac{1}{2}] - [-2^{-n-1}, 2^{-n-1}]$.

For such λ

$$g(\lambda) = 1 - 2^{-n} + 2^{-n-2} - \sum_{k=1}^{n} f_k(\lambda).$$

By (2.4.5) we have

$$-2^{-n-2} \leq \sum_{k=1}^{n+1} f_k(\lambda) - 1 + 2^{-n} - 2^{-n-2} \leq 2^{-n-2} .$$

Whence

$$1 - 2^{-n} \leq \sum_{k=1}^{n+1} f_k(\lambda) \leq 1 - 2^{-n-1} , \quad \lambda \in [-\tfrac{1}{2}, \tfrac{1}{2}] \setminus [-2^{-n-1}, 2^{-n-1}] .$$

For

$$|\lambda| 2^{n+1} \geq 1 , \qquad \lambda \in [-2^{-n-1}, 2^{-n-1}] ,$$

$$g(\lambda) \leq 1 - 2^{-n} + 2^{-n-2} - \sum_{k=1}^{n} f_k(\lambda) .$$

It follows

$$f_{n+1}(\lambda) - \left(1 - 2^{-n} + 2^{-n-2} - \sum_{k=1}^{n} f_k(\lambda)\right) \leq$$

$$\leq f_{n+1}(\lambda) - g(\lambda) \leq 2^{-n-2} .$$

Therefore

$$\sum_{k=1}^{n+1} f_k(\lambda) \leq 1 - 2^{-n-1} , \quad \lambda \in [-2^{-n-1}, 2^{-n-1}]$$

and the relations (2.4.4), (2.4.5) are valid for $n + 1$. By induction we get the validity of (2.4.4), (2.4.5) for any $n \in \mathbb{N}$ and hence $\sum_{k \in \mathbb{N}} f_k(\lambda) = 1$ for all $\lambda \in [-\tfrac{1}{2}, \tfrac{1}{2}] \setminus \{0\}$.

Now let ε_n , $n \in \mathbb{N}$, be the given sequence of numbers. Without loss of generality we assume that $\varepsilon_n \downarrow 0$, $n \to \infty$, $\varepsilon_1 < 1$. Define the family of random variables (2.4.3) taking c_{jk} as in Lemma 2.4.4 and defining the distribution of the random variables ξ_{jk} , $j \in \mathbb{Z}^1$, $k \in \mathbb{N}$, in the following way.

$$P(\xi_{jk} = \pm a_k^{\frac{1}{2}}) = (2a_k)^{-1}, \quad P(\xi_{jk} = 0) = 1 - a_k^{-1} ,$$

where

$$a_k = 2^{k+1} (4N(k) + 2) \varepsilon_{2N(k)}^{-1} , \quad k \in \mathbb{N} .$$

By virtue of (1.4.2) and Lemma 2.4.4 we have

$$\sum_{k \in \mathbb{N}} E X_{nk}^2 = \sum_{k \in \mathbb{N}} \int_{-1/2}^{1/2} f_k(\lambda) d\lambda = 1 .$$

Since the sequence X_{nk} , $k \in \mathbb{N}$, has independent components, this implies that the series $\sum_{k \in \mathbb{N}} X_{nk}$ converges a.s. and also in L_2 . Now it is easy to see that the sequence (2.4.2) is stationary, $E X_n = 0$ and $E X_n^2 < \infty$. Since

$$EX_n X_m = \sum_{k \in \mathbb{N}} EX_{nk} X_{mk} =$$

$$= \sum_{k \in \mathbb{N}} \int_{-1/2}^{1/2} f_k(\lambda) e^{2\pi i |m-n|} d\lambda = \int_{-1/2}^{1/2} e^{2\pi i (m-n)} d\lambda = \delta_{m,n}$$

we conclude that the sequence X_n , $n \in \mathbb{N}$, is orthonormal. Using Lemma 2.4.2   let us estimate the mixing coefficient of this sequence. Let $\alpha_k(n)$ be the mixing coefficient of sequence $X_{n,k}$, $n \in \mathbb{N}$. Then by Lemmas 2.4.1 and 2.4.2 we have

$$\alpha(n) \leq \sum_{k \in \mathbb{N}} \alpha_k(n) \leq \sum_{k : 2N(k) \geq n} \alpha_k(n) \leq \sum_{k : 2N(k) \geq n} a_k^{-1} 2N(k) \leq$$

$$\leq \sum_{k : 2N(k) \geq n} \varepsilon_{2N(k)} \, 2^{-k-2} \leq \varepsilon_n .$$

Thus the constructed process X_n , $n \in \mathbb{N}$, is stationary, orthonormal and has the given α -mixing rate.
Let us verify that for this process the properties 1), 2) and 3) are valid. Put $S_{n,k} = \sum_{j=1}^{n} X_{jk}$. Then for arbitrary n and k we have the representation

$$S_{n,k} = \sum_{j=-N(k)+1}^{N(k)+1} d_j \xi_{j,k} + \sum_{j=-N(k)+n}^{N(k)+n} d_j \xi_{jk} , \qquad (2.4.8)$$

where the coefficients d_j depend on n and k . From (2.4.8) and

$$\sum_{j=-N(k)}^{N(k)} c_{jk} = 0$$

it follows that for $n > 2N(k)+2$

$$P(S_{n,k} \neq 0) \leq \sum_{j=-N(k)+1}^{N(k)+1} P(d_j \xi_{jk} \neq 0) + \sum_{j=-N(k)+n}^{N(k)+n} P(d_j \xi_{jk} \neq 0) \leq \qquad (2.4.9)$$

$$\leq (4N(k)+2) P(\xi_{1,n} \neq 0) = (4N(k)+2) a_k^{-1} \leq 2^{-k-1} .$$

But $P(S_n \neq 0) \leq \sum_{k \in \mathbb{N}} P(S_{n,k} \neq 0) \leq 2^{-1}$ for all $n \in \mathbb{N}$.
Then $P(S_n = 0) \geq \frac{1}{2}$ and the statement 1) is proved.
To prove the property 2) we need the following lemma.

__Lemma 2.4.5.__   Let the spectral density

$$f(\lambda) = \left| \sum_{j=-N}^{N} c_j\, e^{2\pi i j \lambda} \right|^2, \qquad \lambda \in \left[-\tfrac{1}{2}, \tfrac{1}{2}\right] \qquad \text{be such that}$$

$$\sum_{j=-N}^{N} c_j = 0. \qquad\qquad \text{Then}$$

$$\sup_{n \in \mathbb{N}} \int_{-1/2}^{1/2} \frac{\sin^2(\pi \lambda n)}{\sin^2(\pi \lambda)}\, f(\lambda)\, d\lambda < \infty.$$

__Proof.__ It is enough to prove that the relation $f(\lambda)\sin^{-2}(\pi\lambda)$ is bounded. We have

$$\frac{f(\lambda)}{\sin^2(\pi\lambda)} = \left| \frac{\sum_{j=-N}^{N} c_j\, e^{2\pi i j \lambda}}{\sin \pi \lambda} \right|^2 = \left| \frac{\sum_{j=-N}^{N} c_j\, (e^{2\pi i j \lambda} - 1)}{\sin \pi \lambda} \right|^2 \leq$$

$$\leq \left(2 \sum_{j=-N}^{N} |c_j|\,|j|\, \left|\frac{\pi\lambda}{\sin \pi\lambda}\right| \right)^2 \leq \left(2 \sum_{j=-N}^{N} |j\, c_j| \right)^2.$$

Now let $\varepsilon > 0$ be given. Choose such natural $M > 0$ that

$$\sum_{k=M+1}^{\infty} 2^{-k-1} \leq \frac{\varepsilon}{2}.$$

Then (2.4.9) implies

$$\sup_{n \in \mathbb{N}} P\left(\sum_{k=M+1}^{\infty} S_{n,k} \neq 0 \right) \leq \frac{\varepsilon}{2}. \qquad\qquad (2.4.10)$$

By Lemma 2.4.5 for every fixed k , $E\, S_{n,k}^2$ is a bounded function with respect to n . Therefore

$$\sup_{n} E\left(\sum_{k=1}^{M} S_{n,k} \right)^2 < \infty. \qquad\qquad (2.4.11)$$

From (2.4.11) it follows that we can find $a > 0$ such that

$$\sup_{n \in \mathbb{N}} P\left(\left| \sum_{k=1}^{M} S_{n\,k} \right| > a \right) \leq \frac{\varepsilon}{2}. \qquad\qquad (2.4.12)$$

Hence from (2.4.10) and (2.4.12) we obtain

$$\sup_{n \in \mathbb{N}} P(|S_n| > a) \leq \varepsilon.$$

This completes the proof of the Statement 2). The Statement 3)
is the direct consequence of 2).

2.5 Gaussian Random Sequences

The random process ξ_t, $t \in \mathbb{Z}^1$ is called Gaussian if any
random vector ($\xi_{t_1}, \ldots, \xi_{t_n}$), $t_i \in \mathbb{Z}^1$, $i = \overline{1,n}$, has
Gaussian distribution, i.e., its characteristic function is defi-
ned by formula (1.2.1). The stationary Gaussian processes satisfying mixing conditions introduced above always have spectral density. Hence it is a very natural problem to establish the
conditions on the spectral density under which the Gaussian processes possess certain mixing properties (the essential part
of the book /67/ is devoted to this question). For the Gaussian sequences the following statements are valid.

Theorem 2.5.1 (/66/). The Gaussian stationary process satisfies
φ -mixing condition iff it is m -dependent. From this theorem it follows evidently that for Gaussian sequence the definitions of φ and ψ -mixing are equivalent.

Theorem 2.5.2 (/74/). For the Gaussian stationary sequences
the relation

$$\alpha(n) \leq \rho(n) \leq 2\pi \rho(n)$$

is valid.

Now we give the conditions on the spectral density under which
the Gaussian sequence is m -dependent or ρ -mixing (the
proofs of these facts as well as of Theorem 2.5.1 one can
find in the book /66/).

Theorem 2.5.3. The Gaussian stationary sequence is m -de-
pendent iff its spectral density has the form

$$f(\lambda) = |P(e^{i\lambda})|^2,$$

where P is a polynomial with the roots on the unit
circle.

<u>Theorem 2.5.4.</u>   The Gaussian stationary sequence satisfies the ρ -mixing condition iff its spectral density has the form

$$f(\lambda) = |P(e^{i\lambda})|^2 \, \omega(\lambda),$$

where $P(e^{i\lambda})$ is a polynomial with the roots on $|z| = 1$ and for any $\varepsilon > 0$ the function $\omega(\lambda)$ is representable as

$$\omega = exp\{z_\varepsilon + u_\varepsilon + v_\varepsilon\},$$

where z_ε is continuous on C and

$$\|u_\varepsilon\|_\infty + \|v_\varepsilon\|_\infty < \varepsilon$$

($\tilde{v}_\varepsilon$ is the conjugate harmonic to the function v_ε).

From Theorem 2.5.1 it follows easily that if the spectral density of Gaussian sequence is continuous and separated from origin,i.e., $f(\lambda) \geqslant m > 0$ then this sequence satisfies the ρ-mixing condition. Being based on these results it is easy to construct examples of s.r.p. demonstrating a strict imbedding in the relations

$$\prod_{\varphi,\,\frac12 f}^{(k)} \subset \prod_{\alpha,\,\frac14 f}^{(k)}, \qquad k = 0,1,\dots,$$

$$\prod_{\rho,\,f}^{(k)} \subset \prod_{\alpha,\,\frac14 f}^{(k)}, \qquad k = 2,3,\dots$$

2.6 Weak Dependence Conditions for Random Fields. Dobrushin's Example

Defining the weak dependence conditions for random fields ξ_t, $t \in \mathbb{Z}^\nu$, one can use the analogy to the one-dimensional case. Note, however, that for the random fields the direct reformulation of the mixing conditions of 2.2 for instance in the following way:

$$\alpha(n) = \max_{1 \leqslant i \leqslant \nu} \alpha\left(\mathcal{O}_{I_1^{(i)}}, \mathcal{O}_{I_2^{(i)}}\right) \to 0, \qquad n \to \infty, \tag{2.6.1}$$

or

$$\varphi(n) = \max_{1 \leqslant i \leqslant \nu} \varphi\left(\mathcal{O}_{I_1^{(i)}}, \mathcal{O}_{I_2^{(i)}}\right) \to 0, \qquad n \to \infty, \tag{2.6.2}$$

$$I_1^{(i)} = \left\{ t \in \mathbb{Z}^\nu,\ t^{(i)} \leqslant 0 \right\}, \quad I_2^{(i)} = \left\{ t \in \mathbb{Z}^\nu,\ t^{(i)} \geqslant n \right\},$$

$$\mathcal{O}\!t_{I_1^{(i)}} = \mathfrak{S}(\xi_t,\ t \in I_1^{(i)}), \quad \mathcal{O}\!t_{I_2^{(i)}} = \mathfrak{S}(\xi_t,\ t \in I_2^{(i)}), \quad i = \overline{1,\nu}$$

defines the classes of random fields which are very restricted.

Moreover as it shows   Dobrushin's example given below, there are very simple random fields not satisfying the criterion (2.6.2).As to Criterion (2.6.1), Hegerfeldt and Nappy /57/ have shown the existence of some interesting classes of random fields satisfying it.

The weak dependence conditions which are satisfied for large classes of random fields first were introduced by Dobrushin/30/ (see also /84/). The **sense** of these conditions is that their mixing coefficients essentially depend on the subsets of the lattice $\mathbb{Z}^\nu$ involved in its definition. For example,

$$\alpha_I(n) = \sup\left\{ \alpha(\mathcal{O}\!t_I, \mathcal{O}\!t_V),\ V \subset \mathbb{Z}^\nu,\ |\ | < \infty,\ \tau(I,V) \geqslant n \right\} \to 0 ,$$

as $n \to \infty$ and the set $I \subset \mathbb{Z}^\nu,\ |I|$ is fixed,

$$\tau(I,V) = \inf\left\{ \tau(t,s),\ t \in V,\ s \in I \right\},\quad \tau(t,s) = \max_{1 \leqslant i \leqslant \nu} \left| t^{(i)} - s^{(i)} \right|.$$

We will be interested in the following types of mixing conditions for random fields:

$$\alpha(\mathcal{O}\!t_I, \mathcal{O}\!t_V) \leqslant f(I,V)\,\alpha(\tau(I,V)),$$

$$\alpha(\tau) \to 0,\quad \tau \to \infty,$$

where $f\ (.)$ is some function. For example, it is frequently assumed that

$$f(I,V) = |I|^P |V|^S,\quad p,s \in R_+^1 \cup \{0\} \qquad f(I,V) = e^{|I|} \quad \text{and e.c.}$$

Further

$$\alpha_{\mathfrak{m}^1,\mathfrak{m}^2}(n) = \sup\left\{ \alpha(\mathcal{O}\!t_I, \mathcal{O}\!t_V),\ I \in \mathfrak{M}^{(1)},\ V \in \mathfrak{M}^{(2)},\ \tau(I,V) \geqslant n \right\} \to 0,$$

$$n \to \infty \qquad\qquad , \text{ where}$$

$\mathfrak{M}_1,\ \mathfrak{M}_2$ are some classes of subsets of the lattice $\mathbb{Z}^\nu$. Here, for example, one can put

$$\mathfrak{M}^{(i)} = \left\{ V \subset \mathbb{Z}^\nu : \operatorname{diam} V \leqslant d_i \right\},\quad d_i \in R_+^1,\quad i = 1,2,$$

$\operatorname{diam} V$ - is the diameter of the set V, or

$$\mathfrak{m}^{(i)} = \left\{ V \subset \mathbb{Z}^\nu, \; |V| \leq K_i \right\}, \quad K_i \in \mathbb{N}.$$

By exactly the same way the ψ-mixing conditions for random field may be defined.

<u>Dobrushin's Example.</u> Let $\xi_{(t,s)}$, $s, t \in \mathbb{Z}^1$, be the random field such that $\xi_{(t,s)} \in \{0, 1\}$. Suppose that the components of the field are independent for different t and for any fixed $t \in \mathbb{Z}^1$ the random sequence $\xi_{(t,s)}$, $s \in \mathbb{Z}^1$ forms a homogeneous Markov chain with positive transition probabilities not depending on t and not equal to $1/2$. Define

$$A_i^{(n)} = \left\{ \xi_{t0} = i, \; t = 0, 1, \ldots, n \right\}, \quad i = 0, 1.$$

It is easy to see that

$$P\left(\xi_{on} = i_0, \; \xi_{1n} = i_1, \ldots, \xi_{nn} = i_n / A_i^{(n)} \right) = P_{ii_0} P_{ii_1} \ldots P_{ii_n}.$$

Thus the vector ($\xi_{on}, \ldots, \xi_{nn}$) has conditionally independent components. Hence for arbitrary n, $0 < \delta < 1$, $\varepsilon > 0$, one can find a number $N(n)$ such that for $m > N(n)$

$$P\left(\left| \tfrac{1}{m} \sum_{j=1}^{m} \xi_{jn} - P_{11}(n) \right| \leq \varepsilon / A_1^{(n)} \right) \geq 1 - \delta,$$

$$P\left(\left| \tfrac{1}{m} \sum_{j=1}^{m} \xi_{jn} - P_{01}(n) \right| \leq \varepsilon / A_0^{(n)} \right) \leq \delta.$$

These relations show that for such random field the condition (2.6.2) cannot be fulfilled.

2.7 Additions

1) Besides of the mixing coefficients given in this chapter there are frequently used the following measures of dependence between σ - fields:

$$\beta(\mathcal{O}_1, \mathcal{O}_2) = \sup \tfrac{1}{2} \sum_{i=1}^{N} \sum_{j=1}^{M} \left| P(A_i B_j) - P(A_i) P(B_j) \right|$$

(β - mixing condition),

$$I(\mathcal{O}_1, \mathcal{O}_2) = \sup \sum_{i=1}^{N} \sum_{j=1}^{M} P(A_i B_j) \, \ell n \, \frac{P(A_i B_j)}{P(A_i) P(B_j)}$$

(condition of informational regularity).

In both cases the upper bound is taken over all pairs of the partitions $\{A_1, \ldots, A_N\}$, $\{B_1, \ldots, B_M\}$ of the space Ω ,

$$\bigcup_{i=1}^{N} A_i = \bigcup_{j=1}^{M} B_j = \Omega \ , \qquad A_i \cap_{\substack{i\neq j}} A_j = \emptyset, \qquad B_i \cap_{\substack{i\neq j}} B_j = \emptyset .$$

For details see /13/ , /14/ .

2) In Chapter 6 as a special case will be obtained the c.l.th. for s.r.p. satisfying the following condition of weak dependence

for any $k, m \in \mathbb{N}$,

$$\int_{R^{k+m}} \Big| P\Big(\bigcap_{t \in I_k \cup I_n^m} (\xi_t < x_t) \Big) - P\Big(\bigcap_{t \in I_k} (\xi_t < x_t) \Big) P\Big(\bigcap_{t \in I_n^m} (\xi_t < x_t) \Big) \times$$

$$\times \prod_{t \in I_k \cup I_n^m} dx_t \leq \gamma(n), \ n \in \mathbb{N}, \quad \gamma(n) \to 0, \ n \to \infty$$

$$I_k = [-k, 0], \qquad I_n^m = [n, n+m].$$

3) One can use the considered measures of dependence for the definition of weak dependence conditions in the case of non-stationary random process ξ_t, $t \in \mathbb{Z}^1$. For example,

$$\alpha(n) = \sup_{k \in \mathbb{Z}^1} \alpha(\sigma(\xi_s, s \leq k), \sigma(\xi_s, s \geq n+k)).$$

Other mixing conditions are defined by the same way.

4) Give a criterion under which the moving-average process satisfies the α -mixing condition. The formulation of the theorem given below belongs to Gorodetsky /49/ and represents a result of his efforts to correct the work /20/ .

Let ξ_t, $t \in \mathbb{Z}^1$, be the sequence of independent random variables with characteristic functions $\varphi_t(\cdot)$ and the probability densities $p_t(\cdot)$. Take some number sequence $\{g_k\}_{k=0}^{\infty}$, $g_0 \neq 0$, and suppose that there exist the random variables X_j such that the expression

$$X_{jN} = \sum_{i=0}^{N} g_i \xi_{j-i} - X_j$$

tends weakly to zero when $N \to \infty$ and $|X_j| < \infty$ a.s. Define

$$S_i(\delta) = \sum_{j=i}^{\infty} |g_j|^{\delta}, \ \beta(k) = \sum_{i=k}^{\infty} (S_i(\delta))^{1/1+\delta}, \ 0 < \delta < 2,$$

$$\beta(k) = \sum_{i=k}^{\infty} \max\left\{ (S_i(\delta))^{1/1+\delta}, \sqrt{S_i(2)|\ln S_i(2)|} \right\}, \quad \delta \geq 2.$$

__Theorem 2.7.1.__ Let

a) $\displaystyle\int_{-\infty}^{\infty} |p_i(x) - p_i(x+\alpha)|\,dx \leq c_1|\alpha|$,

b) $E|\xi_i|^{\delta} \leq c_2 < \infty$ for some $\delta > 0$ (if $\delta \geq 1$ then it is supposed that $E\xi_i = 0$ and if $\delta \geq 2$ then $D\xi_i = 1$),

c) $\displaystyle q(z) = \sum_{k=0}^{\infty} g_k z^k \neq 0, \quad |z| \leq 1,$

d) $\beta(0) < \infty.$

Then $\left\{X_i\right\}_{i=1}^{\infty}$ satisfies the α -mixing condition with
$$\alpha(n) \leq C_3\,\beta(n), \quad 0 < C_3 < \infty.$$

5) Kesten and O'Braien /73/ have constructed the examples of non-stationary random processes satisfying the j -mixing condition, $j \in (\alpha, \varphi, \psi)$ with given mixing rate.

6) A mixing condition based on characteristic function was considered in /129/ .

7) Heinrich /137/ utilized that Levy's example satisfies indeed the ψ -mixing condition and certain almost-Markov-type mixing condition with an exponential rate in order to show that the arithmetic mean of the continued-fraction expansion $a(t),\ldots,a(t),\ldots$ converges weakly to a stable random variable. An optimal rate of this convergence could be obtained.

3. ASYMPTOTIC BEHAVIOR OF THE VARIANCE, ESTIMATES ON THE MOMENTS AND SOME PROBABILISTIC INEQUALITIES FOR SUMS OF WEAKLY DEPENDENT RANDOM VARIABLES

The estimates and inequalities given below play an important role in the theory of limit theorems for dependent random variables.

3.1 The Variance of the Sum of Random Variables

Let ξ_t , $t \in \mathbb{Z}^\nu$, be a s.r.p. and let $\mathcal{L}_k^m$, $-\infty \le k$, $k \le m \le \infty$, be the linear span of random variables ξ_t, $k \le t \le m$, which is closed in $H(\mathcal{O}_{-\infty}^{+\infty})$. As it was mentioned in Chapter 1, there is a measure preserving transformation T in Ω such that the equality

$$U_\eta(\omega) = \eta(T\omega), \quad \eta \in H(\mathcal{O}_{-\infty}^{\infty})$$

determines in $H(\mathcal{O}_{-\infty}^{\infty})$ an isometric linear operator U the invariance subspace of which is $\mathcal{L}_{-\infty}^{\infty}$.
It is said that s.r.p. ξ_t, $t \in \mathbb{Z}^\nu$, is ergodic if for any $A, B \in \mathcal{O}_{-\infty}^{\infty}$ the following relation is valid:

$$\lim_{n \to \infty} (P(A \cap T^n B) = P(A)P(B), \tag{3.1.1}$$

$$TA = \{\omega : T^{-1}\omega \in A\}, \quad A \in \mathcal{O}_{-\infty}^{\infty} .$$

This definition in terms of the operator U is equivalent with: if $\eta \in H(\mathcal{O}_{-\infty}^{\infty})$ is such that $\eta = U\eta$ then η =const with probability 1.
It is said that s.r.p. ξ_t, $t \in \mathbb{Z}^1$, is regular if

$$\bigwedge_{-\infty < k \le 1} \mathcal{O}_{-\infty}^k = \mathcal{U} ,$$

where the σ-field $\mathcal{U}$ is trivial, i.e., contains only events of probability 0 or 1.
As it has been shown in /66/ the regularity is equivalent with

$$\lim_{n \to -\infty} \sup_{B \in \mathcal{O}_{-\infty}^n} |P(AB) - P(A)P(B)| = 0$$

for each $A \in \mathcal{O}_{-\infty}^{\infty}$.

Hence, α-mixing implies the regularity and the last one implies (3.1.1). Note, that regular s.r.p. has a spectral density and therefore its correlation function.

$$R(n) \to 0 \qquad \text{as} \qquad n \to \infty .$$

Note that the condition $DS_n \to \infty$, $n \to \infty$ is usual in limit theorems for dependent random variables. The following theorem is a starting point in the analysis of the asymptotical behavior of the variance DS_n, $S_n = \sum_{t=1}^{n} \xi_t$ when $n \to \infty$ (the proof can be found in /78/ or /66/).

<u>Theorem 3.1.1 (Leonov)</u>. Let ξ_t, $t \in \mathbb{Z}^1$, be a stationary in a **wide** sense r.p. such that $R(n) \to 0$ as $n \to \infty$. Then either

$$\lim_{n \to \infty} DS_n = \infty \qquad \text{or} \qquad \overline{\lim_{n \to \infty}} DS_n < \infty$$

and the last case holds iff ξ_0 is of the form

$$\xi_0 = \overline{U}g - g, \qquad g \in \mathcal{L}_1^{\infty}, \qquad\qquad (3.1.2)$$

where $\overline{U}$ is the restriction of U on $\mathcal{L}_{-\infty}^{\infty}$.

Thus if ξ_t, $t \in \mathbb{Z}^1$, satisfies the conditions of Theorem 3.1.1 and the equation (3.1.2) is unsolvable, then

$$DS_n \to \infty , \qquad n \to \infty .$$

Gordin /48/ establishes some sufficient conditions of the unsolvability of the equation (3.1.2).

<u>Theorem 3.1.2</u>. Let ξ_t, $t \in \mathbb{Z}^1$, be the ergodic s.r.p. such that $E\xi_0 = 0$, $E\xi_0^2 < \infty$, $R(n) \to 0$, $n \to \infty$ and

$$\mathcal{O}_{-\infty}^{0} \cap \mathcal{O}_1^{\infty} = \mathcal{U} .$$

Then the condition $\lim_{n \to \infty} DS_n < \infty$ implies $\xi_1 = 0$ with probability 1.

<u>Proof</u>. Since $\underline{\lim}_{n \to \infty} DS_n < \infty$ implies $\underline{\lim} DS_n \neq \infty$ then according to Theorem 3.1.1 we have

$$\xi_1 = U\eta - \eta , \qquad \eta \in \mathcal{L}_1^{\infty} .$$

Applying the same theorem to the process ξ_{-t}, $t \in \mathbb{Z}^1$, we obtain

$$\xi_1 = U^{-1}\zeta - \zeta = U(-U^{-1}\zeta) - (-U^{-1}\zeta), \qquad \zeta \in \mathcal{L}_{-\infty}^{1} .$$

Thus both the elements η and $-U^{-1}\eta$ are some solutions of the equation

$$\xi_1 = Uh - h. \tag{3.1.3}$$

As the considered process is ergodic it follows that the difference of any two solutions of (3.1.3) is a constant. Hence

$$\eta = -U^{-1}\zeta + const. \tag{3.1.4}$$

Since η , $U^{-1}\zeta \in \mathcal{L}_{-\infty}^{\infty}$ then $E\eta = EU^{-1}\zeta = 0$ and therefore (3.1.4) implies $\eta = -U^{-1}\zeta$. Hence $\eta \in \mathcal{L}_{-\infty}^{0}$. But at the same time $\eta \in \mathcal{L}_{1}^{\infty}$ and we have

$$\eta \in \mathcal{L}_{-\infty}^{0} \cap \mathcal{L}_{1}^{\infty}. \tag{3.1.5}$$

Using (3.1.5) and the inclusions

$$\mathcal{L}_{-\infty}^{0} \subset H(\mathcal{O}_{-\infty}^{0}), \quad \mathcal{L}_{1}^{\infty} \subset H(\mathcal{O}_{1}^{\infty}).$$

we obtain

$$\eta \in H(\mathcal{O}_{-\infty}^{0}) \cap H(\mathcal{O}_{1}^{\infty}) = H(\mathcal{O}_{-\infty}^{0} \wedge \mathcal{O}_{1}^{\infty}) = H(\mathcal{U}), \tag{3.1.6}$$

i.e., $\eta = 0$ with probability 1. Therefore $\xi_1 = 0$ with probability 1.

<u>Corollary 3.1.1.</u> Let ξ_t, $t \in \mathbb{Z}^1$, be a s.r.p., centred and $E\xi_0^2 < \infty$. If there exists a constant $c, 0 < c < \infty$ such that for any $A \in \mathcal{O}_{-\infty}^{0}$ and any $B \in \mathcal{O}_{1}^{\infty}$

$$P(AB) \geqslant c\, P(A)P(B),$$

then the process satisfies all conditions of Theorem 3.1.2.

<u>Proof.</u> Let us first show that the process ξ_t, $t \in \mathbb{Z}^1$, is regular, i.e.,

$$\bigwedge_{-\infty < k \leqslant 1} \mathcal{O}_{-\infty}^{k} = \mathcal{U}. \tag{3.1.7}$$

Since $\mathcal{O}_{-\infty}^{\infty} = \bigvee_{-\infty < \ell < \infty} \mathcal{O}_{\ell}^{\infty}$, then for each $F \in \bigwedge_{-\infty < k \leqslant 1} \mathcal{O}_{-\infty}^{k}$ and $\varepsilon > 0$ there exist an integer ℓ and an event $F_{\varepsilon} \in \mathcal{O}_{\ell}^{\infty}$ such that $P(F \triangle F_{\varepsilon}) \leqslant \varepsilon$. Here $F \triangle F_{\varepsilon}$ is the symmetric difference of the events F and F_{ε} . From here we can write

$$\varepsilon \geqslant P(\bar{F}F_{\varepsilon}) \geqslant c\, P(\bar{F})P(F_{\varepsilon}) \geqslant c\,(1 - P(F))(P(F) - P(FF_{\varepsilon})) \geqslant$$

$$\geqslant c\,(1 - P(F))(P(F) - \varepsilon), \quad \bar{F} = \Omega \setminus F.$$

Since $\varepsilon > 0$ is arbitrary then (3.1.7) is valid. It remains to show that

$$\mathcal{O}l^{0}_{-\infty} \wedge \mathcal{O}l^{\infty}_{1} = \mathcal{U} .$$

Let $A \in \mathcal{O}l^{0}_{-\infty} \wedge \mathcal{O}l^{\infty}_{1}$, then $0 = P(A\bar{A}) \geqslant c\, P(A) P(\bar{A})$

and Corollary 3.1.1 is proven.

Corollary 3.1.2. Let ξ_t, $t \in \mathbb{Z}^1$ be s.r.p. such that $\psi(1) < 1$

$$E\, \xi_0 = 0 , \quad E\, \xi_0^2 < \infty . \qquad \text{Then}$$

$$DS_n \to \infty , \quad n \to \infty .$$

Proof. Since $\psi(1) < 1$, then for any $A \in \mathcal{O}l^{0}_{-\infty}$, $B \in \mathcal{O}l^{\infty}_{1}$ we have

$$\frac{P(AB)}{P(A)P(B)} - 1 \geqslant -\psi(1), \quad P(AB) \geqslant (1 - \psi(1))\, P(A) P(B) .$$

It remains to put $c = 1 - \psi(1)$ and use Corollary 3.1.1. The following theorems and propositions contain more detailed information about asymptotic behavior of DS_n.

Theorem 3.1.3 (/66/). Let ξ_t, $t \in \mathbb{Z}^1$, be a φ -mixing s.r.p. such that $E\, \xi_0 = 0$, $E\, \xi_0^2 < \infty$. If

$$DS_n \to \infty , \quad n \to \infty ,$$

then

$$DS_n = n\, h(n) ,$$

where $h(n)$ is a slowly varying function on R_+^1.

Remark 3.1.1 (/66/). The conclusion of the theorem 3.1.1 remains valid if

1. $\quad \sigma_n^2 = DS_n \to \infty \qquad \text{as} \quad n \to \infty$

2. for any $\varepsilon > 0$ there exist the integers $p = p(\varepsilon)$ and $N = N(\varepsilon)$ such that

$$\left| E \left(\sum_{j=1}^{n} \xi_j \right) \left(\sum_{j=n+p}^{n+m+p} \xi_j \right) \right| \leqslant \varepsilon \sigma_n \sigma_m , \quad n, m > N(\varepsilon) .$$

Note, that the variance DS_n for α -mixing s.r.p. in general has no such regular behavior (see Chapter 2, Davydov's examples). Nevertheless, we now introduce the conditions under which the regular behaviour of DS_n for α -mixing s.r.p. is preserved.

<u>Theorem 3.1.4.</u>   Suppose ξ_t, $t \in \mathbb{Z}^1$, is s.r.p., $E\xi_0 = 0$ and at least one of the following conditions holds:

1. for some $\delta > 0$, $E|\xi_0|^{2+\delta} < \infty$ and $\sum\limits_{n=1}^{\infty} \alpha^{\delta/2+\delta}(n) < \infty$

2. for some C, $0 < C < \infty$, $|\xi_0| < C$ a.s. and $\sum\limits_{n=1}^{\infty} \alpha(n) < \infty$.

Then $\sigma_0^2 = E\xi_0^2 + 2\sum\limits_{j=1}^{\infty} E\xi_0\xi_j < \infty$. If in addition $\sigma_0^2 \neq 0$ then

$$DS_n = \sigma_0^2 \, n \, (1 + 0(1)), \quad n \to \infty.$$

<u>Proof.</u>
Suppose, for example, that condition 1 holds. By Lemma 2.1.1 we have

$$\sigma_0^2 \leqslant E\xi_0^2 + 2\sum\limits_{j=1}^{\infty} |E\xi_0\xi_j| \leqslant$$

$$\leqslant E\xi_0^2 + 2E^{2/2+\delta}|\xi_0|^{2+\delta} \sum\limits_{n=1}^{\infty} \alpha^{\delta/2+\delta}(n) < \infty.$$

Further

$$DS_n = n E\xi_0^2 + 2\sum\limits_{j=1}^{n} (n-j) E\xi_0\xi_j \leqslant$$

$$\leqslant n\sigma_0^2 - 2\sum\limits_{j=1}^{n} j E\xi_0\xi_j = n\sigma_0^2 \left(1 - \frac{2\sum\limits_{j=1}^{n} j E\xi_0\xi_j}{\sigma_0^2 n}\right).$$

Using the monotonicity of the α-mixing coefficient and condition 1 we have

$$\alpha^{\delta/2+\delta}(j) = 0\,(j^{-1}) \qquad \text{as} \quad j \to \infty.$$

Hence

$$\frac{1}{n}\left|\sum\limits_{j=1}^{n} j E\xi_0\xi_j\right| \leqslant E^{\frac{2}{2+\delta}}|\xi_0|^{2+\delta} \; \frac{\sum\limits_{j=1}^{n} j\,\alpha^{\delta/2+\delta}(j)}{n} \to 0, \quad n \to \infty.$$

Another statement of Theorem 3.1.4 is proved similarly.

<u>Theorem 3.1.5</u> (/28/). Suppose ξ_t, $t \in \mathbb{Z}^1$, is α-mixing s.r.p., $E\xi_0^2 < \infty$ and $DS_n \to \infty$ as $n \to \infty$. Suppose also that sequence $S_n^2 (DS_n)^{-1}$ is uniformly integrable. Then

$$DS_n = nh(n),$$

where $h(n)$ is slowly varying function on R_+^1 .

<u>Proof.</u> It is sufficient to verify the condition 2 of Remark 3.1.1. Let

$$S_{m,n}^p = \sum_{j=n+p}^{n+p+m} \xi_j .$$

By the Schwarz's inequality and Lemma 2.1.1 we have

$$\left| E\left(\frac{S_n}{\sigma_n}\right)\left(\frac{S_{m,n}^p}{\sigma_n}\right)\right| \leq E\left|\frac{S_n}{\sigma_n}\right|\left|\frac{S_{m,n}^p}{\sigma_n}\right| I\left(\frac{S_n}{\sigma_n}\right| > C\right) +$$

$$+ \left| E\left(\frac{S_n}{\sigma_n}\right)\left(\frac{S_{m,n}^p}{\sigma_m}\right) I\left(\left|\frac{S_n}{\sigma_n}\right| \leq C\right) \leq$$

$$\leq E^{1/2}\left|\frac{S_n}{\sigma_n}\right|^2 I\left(\left|\frac{S_n}{\sigma_n}\right| > C\right) + C E^{1/2}\left|\frac{S_m}{\sigma_m}\right|^2 \alpha^{1/2}(p) \leq$$

$$\leq E^{1/2}\left|\frac{S_n}{\sigma_n}\right|^2 I\left(\left|\frac{S_n}{\sigma_n}\right| > C\right) + C\alpha^{1/2}(p) .$$

Now we can choose C and p so that the right side of the last inequality is less than an arbitrary $\varepsilon > 0$.

<u>Theorem 3.1.6</u> (/99/). Let ξ_t, $t \in \mathbb{Z}^1$, be a ρ -mixing s.r.p. $E\xi_0 = 0$, $E\xi_0^2 < \infty$, such that

1) $\sigma_n^2 \to \infty$ as $n \to \infty$.

Then $\sigma_n^2 = n h(n)$, where $h(n)$, $n \in \mathbb{N}$, is a slowly varying function which has a slowly varying extension on R^1 . If in addition we have

2) $\sum_{j=1}^{\infty} \rho(2^j) < \infty$,

then there exists a positive constant $c > 0$ such that

$$\sigma_n^2 = c n (1 + 0(1)), \qquad n \to \infty.$$

<u>Proof.</u> The first conclusion of Theorem 3.1.6 is direct consequence of Lemma 2.1.1 and Remark 3.1.1. Let us prove the second part of Theorem 3.1.6.
Let

45

$$S_n = \sum_{t=1}^{n} \xi_t, \qquad \hat{S}_n = \sum_{t=n+k}^{2n+k} \xi_t, \qquad \bar{S}_\kappa = \sum_{t=2n}^{2n+k} \xi_t.$$

We have

$$E(S_n + \hat{S}_n)^2 = 2ES_n^2 + 2ES_n\hat{S}_n.$$

By Lemma 2.1.1

$$-E^{1/2}S_n^2 E^{1/2}\hat{S}_n^2 \rho(k) \leqslant ES_n\hat{S}_n \leqslant E^{1/2}S_n^2 E^{1/2}\hat{S}^2 \rho(k).$$

Hence

$$2ES_n^2 - 2E^{1/2}S_n^2 E^{1/2}\hat{S}_n^2 \rho(k) \leqslant E(S_n + \hat{S}_n)^2 \leqslant$$
$$\leqslant 2ES_n^2 + 2E^{1/2}S_n^2 E^{1/2}\hat{S}_n^2 \rho(k)$$

and

$$2ES_n^2(1-\rho(k)) \leqslant E(S_n + \hat{S}_n)^2 \leqslant 2ES_n^2(1+\rho(k)).$$

Further

$$ES_{2n}^2 = E(S_n + \hat{S}_n - \bar{S}_\kappa)^2 = E(S_n + \hat{S}_n)^2 - 2E(S_n + \hat{S}_n)\bar{S}_\kappa + E\bar{S}_\kappa^2.$$

From here we obtain

$$2ES_n^2(1-\rho(k))(1-a_{n\kappa}) \leqslant ES_{2n}^2 \leqslant 2ES_n^2(1+\rho(k))(1+a_{n\kappa}), \qquad (3.1.8)$$

where

$$a_{n\kappa} = \frac{|2E(S_n + \hat{S}_n)\bar{S}_\kappa + E\bar{S}_\kappa^2|}{E|S_n + \hat{S}_n|^2} \leqslant$$

$$\leqslant (E|S_n + \hat{S}_n|^2)^{-1}\left(2E^{\frac{1}{2}}(S_n + \hat{S}_n)^2 E^{\frac{1}{2}}|\bar{S}_\kappa|^2 + E\bar{S}_\kappa^2\right) \leqslant$$

$$\leqslant 2\sqrt{E|\bar{S}_\kappa|^2(E|S_n + \hat{S}_n|^2)^{-1}} + E\bar{S}_\kappa^2(E|S_n + \hat{S}_n|^2)^{-1}.$$

Put $\quad n = 2^\nu, \qquad k = 2^{[\frac{\nu}{2}]} \qquad$ and

$$S_{2^p, j} = \sum_{k=2^p(j-1)}^{2^p j} \xi_k, \qquad 0 < p < \nu, \quad j = 1, \ldots, 2^{\nu-p}.$$

Then

$$S_{2^\nu} = \sum_{j=1}^{2^{\nu-p}} S_{2^p, j}.$$

Further by the first part of Theorem 3.1.6 we have

$$\frac{E|\bar{S}_k|^2}{E|S_n+\hat{S}_n|^2} \leq \frac{E|\bar{S}_k|^2}{2E|S_n|^2(1-\rho(k))} = \frac{2^{\left[\frac{\tau}{2}\right]}h(2^{\left[\frac{\tau}{2}\right]})}{2^\tau h(2^\tau)(1-\rho(k))} \qquad (3.1.9)$$

(here τ is such that $\rho(k)<1$).
Using (3.1.8) and the property 4) of a slowly varying function
(see Chapter 1) we conclude that there exists a constant C
such that

$$\frac{E|\bar{S}_k|^2}{E|S_n+\hat{S}_n|^2} \leq C\frac{2^{\frac{\tau}{2}\tau}}{2^\tau} = \frac{C}{2^\tau(1-\frac{\tau}{2})}$$

and therefore $a_{n,k} \leq C\,2^{\tau(\frac{\tau}{2}-1)}$.

By virtue of the inequality (3.1.8) we can write

$$\prod_{j=1}^{\tau-p}(1-\rho(2^{[j/2]}))\prod_{j=1}^{\tau-p}(1-\frac{C}{2^{j/2(1-\tau/2)}})2^{\tau-p}ES^2_{2^p} \leq ES^2_{2^\tau} \leq$$

$$\leq \prod_{j=1}^{\tau-p}(1+\rho(2^{[j/2]}))\prod_{j=1}^{\tau-p}(1+\frac{C}{2^{j/2(1-\tau/2)}})2^{\tau-p}ES^2_{2^p}.$$

Hence
$$\prod_{j=1}^{\tau-p}(1-\rho(2^{[j/2]}))(1-\frac{C}{2^{j/2(1-\tau/2)}}) \leq \frac{h(2^\tau)}{h(2^p)} \leq \prod_{j=1}^{\tau-p}(1+\rho(2^{[j/2]}))(1+\frac{C}{2^{j/2(1-\tau/2)}}).$$

Since $\displaystyle\sum_{j=1}^{\infty}\rho(2^{[j/2]}) < \infty$ and $\displaystyle\sum_{j=1}^{\infty}2^{-j/2(1-\tau/2)} < \infty$

we obtain
$$\lim_{\tau>p>\infty}\frac{h(2^\tau)}{h(2^p)} = 1$$

and therefore $h(2^\tau)$ is convergent to a positive constant. By
the property 3) of a slowly varying function (applied to both of
the functions $h(t)$ and $1/h(t)$) we conclude that $h(n)$ converges to the same constant as $h(2^\tau)$. Theorem 3.1.6 is proven.

<u>Corollary 3.1.3.</u> Let ξ_t, $t\in\mathbb{Z}^1$, be a φ-mixing (ψ-mixing)
s.r.p., $E\xi_0=0$, $E\xi_0^2<\infty$, such that $6_n^2 \to \infty$ as $n\to\infty$
and $$\sum_{j=1}^{\infty}\varphi^{\frac{1}{2}}(2^j) < \infty \ (\ \sum_{j=1}^{\infty}\psi(2^j) < \infty).$$

47

Then there exists a positive constant $C > 0$ such that

$$\sigma_n^2 = C\, n\, (1 + 0\,(1)), \quad n \to \infty.$$

The following theorem gives an estimate for the second term in the asymptotic expansion of the variance.

<u>Theorem 3.1.7</u> (/104/). Let ξ_t, $t \in \mathbb{Z}^1$, be a centered φ -mixing s.r.p. such that for some $\delta > 0$, $E\,|\xi_0|^{2+\delta} < \infty$ and

$$\sum_{n=1}^{\infty} \varphi^{\frac{1}{2}}(n) < \infty. \qquad\qquad (3.1.10)$$

Then

$$DS_n = \sigma_0^2\, n + O\left(n^{\frac{2}{2+\delta}}\right), \quad n \to \infty,$$

where

$$\sigma_0^2 = E\,\xi_0^2 + 2\sum_{k=1}^{\infty} E\,\xi_0\,\xi_k < \infty.$$

<u>Proof</u>. Obviously $\sigma_0^2 < \infty$. It is not difficult to check that

$$DS_n = n\sigma_0^2 - 2n \sum_{k=n}^{\infty} E\,\xi_0\,\xi_k - 2\sum_{k=1}^{n} k\, E\,\xi_0\,\xi_k.$$

Taking into account that (3.1.10) implies $\varphi(k) \leq C_1\, k^{-2}$, and using Lemma 2.2.1 for $p = 2 + \delta$, $q = \frac{2+\delta}{1+\delta}$ we can write

$$\left|\sum_{k=n}^{\infty} E\,\xi_0\,\xi_k\right| \leq C_2 \sum_{k=n}^{\infty} \varphi^{\frac{1+\delta}{2+\delta}}(k) \leq C_3 \sum_{k=n}^{\infty} k^{-\frac{2(1+\delta)}{2+\delta}} \leq$$

$$\leq C_4\, n^{1 - \frac{2(1+\delta)}{2+\delta}} = C_4\, n^{-\frac{\delta}{2+\delta}}.$$

Analogously

$$\sum_{k=1}^{n-1} k\, E\,\xi_0\,\xi_k \leq C_5\, n^{\frac{2}{2+\delta}}$$

and Theorem 3.1.7 is proven.

The following theorem is obtained exactly in the same way.

<u>Theorem 3.1.8</u> (/65/). Let ξ_t, $t \in \mathbb{Z}^1$, be a centered s.r.p., such that

$$\sum_{n=1}^{\infty} n \, | \, \mathrm{Cov} \, (\, \xi_0, \, \xi_n)| < \infty .$$

Then either

$$\sup_n DS_n < \infty \quad \text{or} \quad DS_n = \sigma_0^2 \, n \, (1 + 0(1)), \quad n \to \infty .$$

__Theorem 3.1.9.__ Let ξ_t , $t \in \mathbb{Z}^{\nu}$, $E\xi_0 = 0$ be a s.r.f. such that for some $\delta > 0$, $E \, |\xi_0|^{2+\delta} < \infty$. Then the following statements are valid:

a) If for some $\delta' > 0$, $0 < \delta' < \delta$, the series

$$\sum_{j=1}^{\infty} j^{\nu-1} \, \alpha_{1,1}^{\frac{\delta'}{2+\delta'}} (j)$$

converges, then

$$DS_{I_n} = \sigma_0^2 \, (2n)^{\nu} + O \left(n^{\nu(1-\varepsilon)} \right), \quad \varepsilon = \min \left\{ \frac{1}{2} , \, \frac{(\delta - \delta')}{\delta'(2+\delta)} \right\}$$

$$I_n = [-n, n]^{\nu} , \quad \sigma_0^2 = \sum_{t \in \mathbb{Z}^{\nu}} E \, \xi_0 \, \xi_t < \infty .$$

b) If $\sum\limits_{j=1}^{\infty} j^{\nu-1} \varphi_{1,1}^{\frac{1}{2}} (j) < \infty$ then $DS_{I_n} = \sigma_0^2 \, (2n)^{\nu} + O \left(n^{\frac{2\nu}{2+\delta}} \right).$

__Proof.__ Let us prove the statement a). We have

$$DS_{I_n} = E \, \xi_0^2 \cdot (2n)^{\nu} + \sum_{t \in I_n} \sum_{k \in I_n \setminus \{t\}} E \, \xi_t \, \xi_k =$$

$$= (2n)^{\nu} \left(E \, \xi_0^2 + \sum_{t \in \mathbb{Z}^{\nu} \setminus \{0\}} E \, \xi_0 \, \xi_t \right) + \sum_{t \in I_n} \sum_{k \in I_n \setminus \{t\}} E \, \xi_t \, \xi_k \, -$$

$$- (2n)^{\nu} \sum_{t \in I_n \setminus \{0\}} E \, \xi_0 \, \xi_t - (2n)^{\nu} \sum_{\substack{t \notin I_n \setminus \{0\} \\ t \in \mathbb{Z}^{\nu} \setminus \{0\}}} E \, \xi_0 \, \xi_t .$$

Further let us estimate the last term. Using Lemma 2.2.1 we can write

$$\sum_{\substack{t \notin I_n \setminus \{0\} \\ t \in \mathbb{Z}^{\nu} \setminus \{0\}}} E \, \xi_0 \, \xi_t \leq C_1 \sum_{j \geq n} j^{\nu-1} \, \alpha_{1,1}^{\frac{\delta}{2+\delta}} (j) \leq$$

$$\leq C_2 \sum_{j \geq n} j^{\nu-1} \left(\frac{1}{j^{\nu}} \right)^{\frac{\delta}{2+\delta} \cdot \frac{2+\delta}{\delta'}} = C_2 \sum_{j \geq n} \frac{1}{j^{1 + \frac{2\nu(\delta - \delta')}{\delta(2+\delta)}}} \leq$$

$$\leq C_3 \, n^{\frac{2\nu(\delta' - \delta)}{\delta'(2+\delta)}} .$$

We find

$$\Big| \sum_{t\in I_n} \sum_{k\in I_n\setminus\{t\}} E\xi_t\xi_k - (2n)^\nu \sum_{k\in I_n\setminus\{0\}} E\xi_0\xi_k \Big| =$$

$$=\Big| \sum_{t\in I_n} \sum_{k\in I_n\setminus\{t\}} E\xi_t\xi_k - \sum_{t\in I_n} \sum_{k\in I_n\setminus\{0\}} E\xi_0\xi_k \Big| =$$

$$=\Big| \sum_{t\in I_n\setminus\{0\}} \Big(\sum_{k\in I_n\setminus\{t\}} E\xi_0\xi_{k-t} - \sum_{k\in I_n\setminus\{0\}} E\xi_0\xi_k \Big) \Big| =$$

$$= \sum_{t\in I_n\setminus\{0\}} \Big| \sum_{k:\,k+t\in I_n} E\xi_0\xi_k - \sum_{k\in I_n\setminus\{0\}} E\xi_0\xi_k \Big| \leq$$

$$\leq \sum_{t\in I_n\setminus\{0\}} \sum_{\substack{k:\,k+t\in I_n \\ k\notin I_n\setminus\{0\}}} |E\xi_0\xi_k| \leq \sum_{k\in I_n\setminus\{0\}} \Big| n^\nu - \prod_{j=1}^{\nu}(n-k^{(j)}) \Big| \, |E\xi_0\xi_k| \leq$$

$$\leq C\, n^{\nu-1} \sum_{j=1}^{n} j^\nu \alpha_{1,1}^{\frac{\delta}{2+\delta}}(j).$$

Finally

$$n^{-1} \sum_{j=1}^{n} j^\nu \alpha_{1,1}^{\frac{\delta}{2+\delta}}(j) \leq \frac{C_1}{\sqrt{n}} + \frac{\sum_{j\geq\sqrt{n}}^{n} j^\nu \alpha_{1,1}^{\frac{\delta}{2+\delta}}(j)}{n} \leq$$

$$\leq \frac{C_1}{\sqrt{n}} + \frac{1}{n} \sum_{j\geq\sqrt{n}}^{n} j^\nu \Big(\frac{1}{j^\nu}\Big)^{\frac{\delta}{2+\delta}\frac{2+\delta'}{\delta'}} \leq \frac{C_1}{\sqrt{n}} + \sum_{j\geq\sqrt{n}}^{\infty} j^{2\nu\frac{\delta'-\delta}{(2+\delta)\delta'}-1} \leq$$

$$\leq \frac{C_1}{\sqrt{n}} + \frac{C_2}{n^{\frac{\gamma(\delta-\delta')}{(2+\delta)\delta'}}} \, .$$

The statement b) is proved similarly.

Theorem 3.1.10. Let a s.r.p. ξ_t, $t \in \mathbb{Z}^1$, be ρ-mixing centered, $E\xi_0^2 < \infty$ and $\sigma_n^2 \to \infty$, as $n \to \infty$. If for some $\varepsilon > 0$ $\lim_{n \to \infty} R(n) n^{1-\varepsilon}$ exists, then its value is equal to zero.

Proof. By Theorem 3.1.6 $\sigma_n^2 = n\,h(n)$, where $h(n)$ is a slowly varying function. Hence

$$h(n) = \frac{2}{n} \sum_{j=1}^{n} (n-j)\, R(j) + R(0)$$

and also for any $\varepsilon > 0$

$$\lim_{n \to \infty} h(n) n^{-\varepsilon} = 0 \, . \qquad (3.1.11)$$

It is well known that for two number sequences x_n, $n \in \mathbb{N}$, and y_n, $n \in \mathbb{N}$, $y_{n+1} > y_n$, $n \in \mathbb{N}$,

$$\lim_{n \to \infty} \frac{x_n}{y_n} = \lim_{n \to \infty} \frac{x_n - x_{n-1}}{y_n - y_{n-1}}$$

(if the last limit exists). Using this fact and (3.1.11) we can write

$$\lim_{n \to \infty} \frac{h(n)}{n^\varepsilon} = 2 \lim_{n \to \infty} \frac{\sum_{j=1}^{n} (n-j)\, R(j) - \sum_{j=1}^{n-1} (n-1-j)\, R(j)}{n^{1+\varepsilon} - (n-1)^{1+\varepsilon}}$$

$$= \frac{2}{1+\varepsilon} \lim_{n \to \infty} \frac{\sum_{j=1}^{n} R(j)}{n^\varepsilon} = \frac{2}{(1+\varepsilon)\varepsilon} \lim_{n \to \infty} n^{1-\varepsilon} R(n) = 0 \, .$$

This criterion may be used to construct various examples of ρ-mixing s.r.p. not satisfying the strong mixing conditions.

3.2 Estimates on Moments of Sums of Weakly Dependent

Random Variables

The important technical lemma given below essentially is contained in /65/.

Lemma 3.2.1. Let a s.r.p. ξ_t, $t \in \mathbb{Z}^1$ be such that :

1. for some $\delta > 0$, $E|\xi_0|^{2+\delta} < \infty$;

2. $\qquad \sigma_n^2 = n h(n)$, $\qquad n \in \mathbb{N}$, $\qquad$ where $h(n)$

is slowly varying function on R^1;

3. for any $\varepsilon > 0$ there exists $n_0 \in \mathbb{N}$ such that for $n > n_0$ the following inequality holds:

$$a_{2n} \leqslant (2+\varepsilon) a_n + C (\sigma_n^{2+\delta})^p a_n^q , \quad p > 0, q \geqslant 0, \ p+q = 1 . \tag{3.2.1}$$

Then $\qquad a_n \leqslant CE|S_n|^{2+\delta}, \quad 0 < C < \infty .$

<u>Proof.</u> $\qquad$ Let $\beta_n = a_n \sigma_n^{-(2+\delta)}$. Using (3.2.1) we can write

$$\beta_{2n} \leqslant (2+\varepsilon)\left(\frac{\sigma_n}{\sigma_{2n}}\right)^{2+\delta} \beta_n + C\left(\frac{\sigma_n}{\sigma_{2n}}\right)^{2+\delta} \beta_n^q .$$

Since

$$\frac{\sigma_{2n}^2}{\sigma_n^2} = \frac{2n h(n)}{n h(n)} \to 2 , \qquad n \to \infty,$$

then for any ε_0 and sufficiently large n we have

$$\left(\frac{\sigma_n}{\sigma_{2n}}\right)^{2+\delta} < 1 + \varepsilon_0 .$$

Hence for each $\varepsilon_1 > 0$ and sufficiently large n the following inequality is valid:

$$\beta_{2n} \leqslant (1+\varepsilon_1) 2^{-\frac{\delta}{2}} \beta_n + C_1 \beta_n^q , \quad 0 < C_1 < \infty . \tag{3.2.1'}$$

Suppose that ε_1 and the constant C_2 are such that

$$\lambda = (1+\varepsilon_1) 2^{-\frac{\delta}{2}} + C_1 C_2^{-p} < 1 \qquad \text{and}$$

$$\mathbb{N}_1 = \left\{ n \in \mathbb{N} : \beta_n < C_2 \right\} . \qquad \text{Then for } n \in \mathbb{N}_1$$

(3.2.1)' implies $\beta_{2n} < C_3$. On the other hand for $n \in \mathbb{N} \setminus \mathbb{N}_1$ we have $\beta_{2n} \leqslant \lambda \beta_n$. Finally

$$\beta_{2n} \leqslant \lambda \beta_n + C_3 , \qquad 0 < C_3 < \infty . \tag{3.2.2}$$

From (3.2.2) we directly obtain

$$\sup_{\tau} \beta_{2^\tau} = \sup_{\tau} \frac{a_{2^\tau}}{\sigma_{2^\tau}^{2+\delta}} = C_4 < \infty .$$

Hence

$$a_{2^\tau} \leqslant C_4 \sigma_{2^\tau} , \qquad \tau \in \mathbb{N} . \tag{3.2.3}$$

Now suppose that $n = \nu_0 2^{\tau} + \nu_1 2^{\tau-1} + \ldots + \nu_{\tau}$, where $\nu_0 = 1$ and $\nu_j \in \{0, 1\}$, $j > 0$. Let us represent the sum S_n in the form

$$S_n = \sum_{j=1}^{n} \xi_j = \left(\xi_1 + \cdots + \xi_{i_1} \right) + \left(\xi_{i_1+1} + \cdots + \xi_{i_2} \right) + \cdots + \left(\xi_{i_{\tau}+1} + \cdots + \xi_n \right),$$

where the number of summands in j-terms is equal to $\nu_j 2^j$, $j = 0, \ldots, \tau$. By Minkovski's inequality and (3.2.3) we can write

$$a_n \leq \left| \sum_{j=0}^{\tau} E^{\frac{1}{2+\delta}} \left| \xi_1 + \cdots + \xi_{\nu_j 2^{\tau-j}} \right|^{2+\delta} \right|^{2+\delta} \leq$$

$$\leq C_4 \left(\sum_{j=0}^{\tau} \sigma_{2^{\tau-j}} \right)^{2+\delta} = C_4 \sigma_n^{2+\delta} \left(\sum_{j=0}^{\tau} \frac{\sigma_{2^{\tau-j}}}{\sigma_n} \right)^{2+\delta} =$$

$$= C_4 \sigma_n^{2+\delta} \left(\sum_{j=0}^{\tau} 2^{\frac{\tau-j}{2} - \frac{n}{2}} \frac{h(2^{\tau-j})}{h(2^{\tau})} \cdot \frac{h(2^{\tau})}{h(n)} \right)^{2+\delta}.$$

Using the following property of slowly varying function

$$h = \sup_{\tau} \sup_{2^{\tau} \leq n < 2^{\tau+1}} \frac{h(2^{\tau})}{h(n)} < \infty$$

we have

$$a_n \leq C_5 \sigma_n^{2+\delta} \left(\sum_{j=0}^{\tau} 2^{-\frac{j}{2}} \frac{h(2^{\tau-j})}{h(2^{\tau})} \right)^{2+\delta}.$$

Now for any $\varepsilon_2 > 0$ there exists the integer n_0 such that

$$\frac{h(n)}{h(2n)} < 1 + \varepsilon_2, \quad n > n_0.$$

For each $j \in [2, \tau]$ we have

$$\frac{h(2^{\tau-j})}{h(2^{\tau})} = \left(\frac{h(2^{\tau-1})}{h(2^{\tau})} \frac{h(2^{\tau-2})}{h(2^{\tau-1})} \ldots \frac{h(2^{\tau-s})}{h(2^{\tau-s-1})} \right) \left(\frac{h(2^{\tau-s-1})}{h(2^{\tau-s})} \ldots \frac{h(2^{\tau-j})}{h(2^{\tau-j-1})} \right).$$

Choose some $s \in \mathbb{N}$ such that $2^{\tau-s} \geq n_0$, $2^{\tau-s-1} < n_0$. Then

$$\frac{h(2^{\tau-j})}{h(2^{\tau})} \le C_6 (1+\varepsilon_2)^{s-1} \le C_6 (1+\varepsilon_2)^{j-1}.$$

From here

$$a_n \le C_7 \, 6_n^{2+\delta} \Big(\sum_{j=0}^{\tau} \Big(\frac{1+\varepsilon_2}{\sqrt{2}} \Big)^j \Big)^{2+\delta}.$$

Now if ε_2 is such that $\dfrac{1+\varepsilon_2}{\sqrt{2}} < 1$, then

$$a_n \le C_8 \, 6_n^{2+\delta}$$

which completes the proof of Lemma 3.2.1.

<u>Proposition 3.2.1.</u> Let ξ_t, $t \in \mathbb{Z}^1$, be the α -mixing s.r.p. and for some δ, $0 < \delta < 1$, $\sum_{j=1}^{\infty} \alpha^{\frac{\delta}{2+\delta}}(j) < \infty$, $E\xi_0 = 0$. Suppose also that $|\xi_t| < C$ with probability 1 and

$$6_0^2 = E(\xi_0)^2 + 2\sum_{j=1}^{\infty} E\,\xi_0\,\xi_j > 0.$$

Then

$$a_n = E\,|S_n|^{2+\delta} \le C_1 \, 6_n^{2+\delta}, \qquad 0 < C_1 < \infty .$$

<u>Proof.</u> Let us first show that for any $\varepsilon > 0$ and function $k = k(n) = [n^{\gamma}]$, $\frac{\delta}{2} < \gamma < \frac{2+\delta}{2(1+\delta)}$ one can choose the constant $C_2 > 0$ such that for sufficiently large n

$$E\,|S_n + \hat{S}_n|^{2+\delta} \le (2+\varepsilon)a_n + C_2 \, 6_n^{2+\delta}, \qquad (3.2.4)$$

where

$$\hat{S}_n = \sum_{j=n+k+1}^{2n+k} \xi_j .$$

We can write

$$E\,|S_n + \hat{S}_n|^{2+\delta} \le E\,|S_n + \hat{S}_n|^2 \big(|S_n|^{\delta} + |\hat{S}_n|^{\delta} \big) \le 2a_n +$$

$$+ 2E\,|S_n|^{1+\delta}|\hat{S}_n| + 2E\,|S_n||\hat{S}_n|^{1+\delta} + E\,|S_n|^2|\hat{S}_n|^{\delta} + E\,|S_n|^{\delta}|\hat{S}_n|^2.$$

Using Lemma 2.1.1 we have

$$E\,|S_n||\hat{S}_n|^{1+\delta} \le C_3 n^{\delta} E\,|S_n||\hat{S}_n| \le C_3 n^{\delta} E^{\frac{1}{2+\delta}}|S_n| \, E^{\frac{1}{2+\delta}}|\hat{S}_n| \, \alpha^{\frac{\delta}{2+\delta}}(k) +$$

$$+ C_4 \, 6_n^{2+\delta} = C_3 n^{\delta} E^{\frac{2}{2+\delta}}|S_n|^{2+\delta} \alpha^{\frac{\delta}{2+\delta}}(k) + C_4 \, 6_n^{2+\delta} .$$

In exactly the same way

$$E\,|S_n|^{1+\delta}\,|\hat{S}_n| \le C_3\,n^\delta\,E^{\frac{2}{2+\delta}}|S_n|^{2+\delta}\,\alpha^{\frac{\delta}{2+\delta}}(k)+C_4\,\delta_n^{2+\delta}\,.$$

Further

$$E\,|S_n|^2\,|\hat{S}_n|^\delta \le C_3\,n^\delta\,E\,|S_n|^{2-\delta}\,|\hat{S}_n|^\delta \le C_3\,n^\delta\,E^{\frac{2-\delta}{2+\delta}}|S_n|^{2+\delta}\times$$

$$\times\,E^{\frac{\delta}{2+\delta}}|\hat{S}_n|^{2+\delta}\,\alpha^{\frac{\delta}{2+\delta}}(k)+C_5\,\delta_n^{2+\delta} \le C_3\,n^\delta\,E^{\frac{2}{2+\delta}}|S_n|^{2+\delta}\,\alpha^{\frac{\delta}{2+\delta}}(k)+C_5\,\delta_n^{2+\delta}\,.$$

Analogously

$$E\,|S_n+\hat{S}_n| \le 2E\,|S_n|^{2+\delta}+C_6\,n^\delta\,E^{\frac{2}{2+\delta}}|S_n|^{2+\delta}\,\alpha^{\frac{\delta}{2+\delta}}(k)+C_7\,\delta_n^{2+\delta}\,.$$

Hence

$$E\,|S_n+\hat{S}_n| \le 2E\,|S_n|^{2+\delta}+C_6\,n^\delta\,E^{\frac{2}{2+\delta}}|S_n|^{2+\delta}\,\alpha^{\frac{\delta}{2+\delta}}(k)+C_7\,\delta_n^{2+\delta}\,.$$

Since

$$\frac{n^\delta\,E^{\frac{2}{2+\delta}}|S_n|^{2+\delta}\,\alpha^{\frac{\delta}{2+\delta}}(k)}{E\,|S_n|^{2+\delta}} \le \frac{C_8\,n^\delta}{k(n)\,E^{\frac{\delta}{2+\delta}}|S_n|^{2+\delta}} \le \frac{C_8\,n^{\frac{\delta}{2}}}{n^\gamma}$$

and $\gamma > \frac{\delta}{2}$ then the estimate (3.2.4) holds.

By Minkovski's inequality we can write

$$a_{2n} = E\,\Big|S_n+\hat{S}_n+\sum_{j=n+1}^{n+k}\xi_j-\sum_{j=2n+1}^{2n+k}\xi_j\Big|^{2+\delta} \le$$

$$\le E\,|S_n+\hat{S}_n|^{2+\delta}\Big(1+\frac{2\,a_k^{\frac{1}{2+\delta}}}{E^{\frac{1}{2+\delta}}|S_n+\hat{S}_n|^{2+\delta}}\Big)^{2+\delta}\,. \tag{3.2.5}$$

Since $|\xi_t| < C$ then using Lyapunov's inequality we conclude that

$$\frac{a_k}{E\,|S_n+\hat{S}_n|^{2+\delta}} \le \frac{C_9\,k^{1+\delta}}{n^{\frac{2+\delta}{2}}} = \frac{C_9\,n^{\gamma(1+\delta)}}{n^{\frac{2+\delta}{2}}}\,.$$

But $\gamma < \frac{2+\delta}{2(1+\delta)}$ and hence

$$\lim_{n\to\infty}\frac{a_k}{E\,|S_n+\hat{S}_n|^{2+\delta}} = 0\,. \tag{3.2.6}$$

From (3.2.6), (3.2.5) and (3.2.4) we obtain that for sufficiently large n the relation (3.2.1) is valid for $p = 1$,
$q = 0$.

$\underline{\text{Lemma 3.2.2}}$ ($\underline{/65/}$). Suppose that s.r.p. ξ_t, $t \in \mathbb{Z}^1$, $E\xi_0 = 0$, $E\xi_0^2 < \infty$ is ρ-mixing and

$$\sigma_n^2 \to \infty \qquad \text{as} \quad n \to \infty.$$

Then

$$a_n \leqslant C\,\sigma_n^{2+\delta}, \qquad 0 < C < \infty, \quad 0 < \delta < 1.$$

$\underline{\text{Proof.}}$ Let $\hat{S}_n = \sum\limits_{j=n+k+1}^{2n+k} \xi_j$, where k is a constant which we will choose later. We have

$$E\,|S_n + \hat{S}_n|^{2+\delta} \leqslant 2\,a_n + E\,|S_n|^2\,|\hat{S}_n|^\delta + E\,|\hat{S}_n|^2\,|S_n|^\delta + \qquad (3.2.7)$$
$$+ 2E\,|\hat{S}_n|^{1+\delta}\,|S_n| + 2E\,|S_n|^{1+\delta}\,|\hat{S}_n| .$$

By Hölder's inequality and Lemma 2.1.1 we can write

$$E\,|S_n|^2\,|\hat{S}_n|^\delta \leqslant E^{\frac{2-\delta}{2+\delta}}\,|S_n|^{2+\delta}\,E^{\frac{2\delta}{2+\delta}}\,|S_n|^{\frac{2+\delta}{2}}\,|\hat{S}_n|^{\frac{2+\delta}{2}} \leqslant$$
$$\leqslant \sigma_n^{2\delta}\,a_n^{\frac{2-\delta}{2+\delta}} + \rho^{\frac{2\delta}{2+\delta}}(k)\,a_n .$$

Similarly

$$E\,|S_n|^{1+\delta}\,|\hat{S}_n| \leqslant E^{\frac{\delta}{2+\delta}}\,|S_n|^{2+\delta}\,E^{\frac{2}{2+\delta}}\,|S_n|^{\frac{2+\delta}{2}}\,|\hat{S}_n|^{\frac{2+\delta}{2}} \leqslant$$
$$\leqslant a_n^{\frac{\delta}{2+\delta}}\,\sigma_n^2 + \rho^{\frac{2}{2+\delta}}(k)\,a_n \leqslant \sigma_n^{2\delta}\,a_n^{\frac{2-\delta}{2+\delta}} + \rho^{\frac{2\delta}{2+\delta}}(k)\,a_n .$$

Estimating the other summands in (3.2.7) just in the same way we obtain

$$E\,|S_n + \hat{S}_n|^{2+\delta} \leqslant \left(2 + 6\rho^{\frac{2\delta}{2+\delta}}(k)\right) a_n + 6\,\sigma_n^{2\delta}\,a_n^{\frac{2-\delta}{2+\delta}} .$$

Hence for any $\varepsilon > 0$ there exists an integer k such that for sufficiently large n the following inequality is valid

$$E\,|S_n + \hat{S}_n|^{2+\delta} \leqslant (2+\varepsilon)\,a_n + 6\,\sigma_n^{2\delta}\,a_n^{\frac{2-\delta}{2+\delta}} .$$

Hence

$$a_{2n} \leqslant (2+\varepsilon)\,a_n + 7\,\sigma_n^{2\delta}\,a_n^{\frac{2-\delta}{2+\delta}} .$$

It remains to use the Lemma 3.2.1.
As we have indicated

$$\rho(n) \leq 2\varphi^{\frac{1}{2}}(n)$$

and therefore the following corollary holds.

__Corollary 3.2.1.__ Let $\xi_t, t \in \mathbb{Z}^1$, $E\xi_t = 0$, be φ-mixing s.r.p. and for some δ, $0 < \delta < 1$, $E|\xi_0|^{2+\delta} < \infty$. If

$$\sigma_n^2 \to \infty, \quad n \to \infty ,$$

then

$$a_n \leq C\sigma_n^{2+\delta}, \quad 0 < C < \infty .$$

For ψ-mixing s.r.p. the following lemma is valid.

__Lemma 3.2.3 (/86/).__ Let ξ_t, $t \in \mathbb{Z}^1$, $E\xi_0 = 0$, $E\xi_0^2 < \infty$, be the ψ-mixing s.r.p. and $\sigma_n^2 \to \infty$, $n \to \infty$.
Then there exists a constant $C < \infty$ such that

$$E|S_n|^2|\hat{S}_n| \leq C\sigma_n^3, \quad \hat{S}_n = \sum_{\substack{s,t=1 \\ s \neq t}}^{n} \xi_t \xi_s .$$

__Proof.__ Note that by Lemma 2.1.1 the moment $E|S_n|^2|\hat{S}_n|$ exists. We have

$$a_{2n} = E\left|\sum_{\substack{t,s=1 \\ t \neq s}}^{2n} \xi_t \xi_s\right|\left|\sum_{t=1}^{2n} \xi_t\right| = E\left|\left(\sum_{\substack{t,s=1 \\ t \neq s}}^{n} \xi_t \xi_s\right)\left(\sum_{t=1}^{n} \xi_t\right) + \right.$$

$$+ 2\left(\sum_{t=1}^{n} \xi_t\right)^2\left(\sum_{t=n+1}^{2n} \xi_t\right) + 2\left(\sum_{t=1}^{n} \xi_t\right)\left(\sum_{t=n+1}^{2n} \xi_t\right)^2 +$$

$$+ \left(\sum_{\substack{t,s=n+1 \\ t \neq s}}^{2n} \xi_t \xi_s\right)\left(\sum_{t=1}^{n} \xi_t\right) + \left(\sum_{\substack{t,s=1 \\ t \neq s}}^{n} \xi_t \xi_s\right)\left(\sum_{t=n+1}^{2n} \xi_t\right) +$$

$$+ \left(\sum_{t,s=n+1}^{2n} \xi_t \xi_s\right)\left(\sum_{t=n+1}^{2n} \xi_t\right)\right| \leq 2a_p +$$

$$+ 2E\left|\sum_{t=1}^{n} \xi_t\right|^2\left|\sum_{t=n+1}^{2n} \xi_t\right| + 2E\left(\left|\sum_{t=1}^{n} \xi_t\right|\left|\sum_{t=n+1}^{2n} \xi_t\right|^2\right) +$$

$$+ E\left(\sum_{\substack{t,s=n+1 \\ t \neq s}}^{2n} \xi_t \xi_s\right)\left(\sum_{t=1}^{n} \xi_t\right) + E\left|\left(\sum_{t,s=1}^{n} \xi_t \xi_s\right)\left(\sum_{t=p+1}^{2n} \xi_t\right)\right| .$$

Let us estimate each summand in (3.2.8). Using Lemma 2.1.1 we have

$$E\left|\sum_{t=1}^{n}\xi_t\right|^2\left|\sum_{t=n+1}^{2n}\xi_t\right|\leq(1+\psi(1))E\left|\sum_{t=1}^{n}\xi_t\right|^2 E\left|\sum_{t=n+1}^{2n}\xi_t\right|\leq$$

$$\leq(1+\psi(1))\,\sigma_n^3.$$

Analogously

$$E\left|\sum_{t=1}^{n}\xi_t\right|\left|\sum_{t=n+1}^{2n}\xi_t\right|^2\leq(1+\psi(1))\,\sigma_n^3.$$

Further

$$E\left(\left|\sum_{t,s=n+1}^{2n}\xi_t\xi_s\right|\left|\sum_{t=1}^{n}\xi_t\right|\right)\leq E\left|\sum_{t=n+1}^{2n}\xi_t\right|^2\left|\sum_{t=1}^{n}\xi_t\right|+$$

$$+\sum_{t=n+1}^{2n}E\,\xi_t^2\left|\sum_{t=1}^{n}\xi_t\right|\leq(1+\psi(1))\sigma_n^3+(1+\psi(1))\sum_{t=n+1}^{2n}E\,\xi_t^2,$$

$$E\left|\sum_{t=1}^{n}\xi_t\right|\leq(1+\psi(1))\left(\sigma_p^3+p\sigma_p\right)\leq C_1\sigma_n^3,\quad 0<C_1<\infty,$$

the last summand in (3.2.8) is estimated in just the same way as the previous members. Finally we have

$$a_{2n}\leq 2a_n+C_2\,\sigma_n^3.$$

3.3 Some Probabilistic Inequalities

Below we will introduce some probabilistic inequalities which represent the generalization of well-known inequalities of Ottaviani and Hoffman-Jorgensen. The first two inequalities can be deduced from Billingsley (1968 p.175) and are due to Oodaira and Yoshihara /87/ and Peligrad /100/ , respectively. The inequality (3.3.3) must be attributed to Peligrad

__Theorem 3.3.1.__ Let X_1, X_2,...., X_n be the random variables and let $S_i=\sum_{j=1}^{i}X_j$, $i=\overline{1,n}$, $k=[\frac{n}{p}]$, $p\in\mathbb{N}$. If

$$k\geq2,\quad\max_{0\leq i\leq k-2}P\left(|S_n-S_{(i+2)p}|\geq\tfrac{\lambda}{3}\right)\leq\eta<1$$

then the following inequality holds:

$$P\left(\max_{1\leq i\leq n}|S_i|\geq\lambda\right)\leq\frac{1}{1-\eta}\left(P(|S_n|\geq\tfrac{\lambda}{3})+k\alpha(p)+2\tfrac{n}{p}P(\sum_{j=1}^{2p}|X_j|>\tfrac{\lambda}{3})\right)\tag{3.3.1}$$

__Proof.__ Let $E_i = \{\max_{j<i} |S_j| < \lambda \le |S_i|\}$, $i = \overline{1,n}$.
Then we have

$$P(\max_{1\le i\le n} |S_i| \ge \lambda) = P(\bigcup_{j=1}^{n} (E_j \cap \{|S_j| \ge \lambda\})) \le$$

$$\le P(|S_n| \ge \tfrac{\lambda}{3}) + P(\bigcup_{j=1}^{n} (E_j \cap \{|S_n - S_j| \ge \tfrac{2\lambda}{3}\})) \le$$

$$\le P(|S_n| \ge \tfrac{\lambda}{3}) + \sum_{i=0}^{k-2} P((\bigcup_{j=1}^{P} E_{iP+j}) \cap \{|S_n - S_{(i+2)P}| \ge \tfrac{\lambda}{3}\}) +$$

$$+ \sum_{i=0}^{k-2} P(\bigcup_{j=1}^{P} (E_{iP+j} \cap \{|S_{(i+2)P} - S_{iP+j}| \ge \tfrac{\lambda}{3}\})) +$$

$$+ \sum_{j=(k-1)P+1}^{n} P(E_j \cap \{|S_n - S_j| \ge \tfrac{2\lambda}{3}\}) \le$$

$$\le P(|S_n| \ge \tfrac{\lambda}{3}) + \sum_{i=0}^{k-2} P(\bigcup_{j=1}^{P} E_{iP+j}) P(|S_n - S_{(i+2)P}| \ge \tfrac{\lambda}{3}) + k\alpha(P)$$

$$+ 2[\tfrac{n}{P}] P(\sum_{j=1}^{2P} |X_j| \ge \tfrac{\lambda}{3}) \le$$

$$\le P(|S_n| \ge \tfrac{\lambda}{3}) + \max_{0\le i\le k-2} P(|S_n - S_{(i+2)P}| \ge \tfrac{\lambda}{3}) P(\max_{1\le i\le n} |S_i| \ge \lambda) +$$

$$+ k\alpha(P) + 2[\tfrac{n}{P}] P(\sum_{j=1}^{2P} |X_j| \ge \tfrac{\lambda}{3}).$$

This implies the inequality (3.3.1).

__Theorem 3.3.2.__ (Generalized Ottaviani's inequality)
Let $X_1, X_2, \ldots, X_n$ be the random variables and suppose that
for some $\lambda > 0$, $P \in \mathbb{N}$,

$$\varphi(P) + \max_{0\le i\le n} P(|S_n - S_i| \ge \tfrac{\lambda}{3}) \le \eta < 1.$$

Then the following relation holds

$$P(\max_{1\le i\le n} |S_i| \ge \lambda) \le$$

$$\leq \frac{1}{1-\eta} P\left(|S_n| > \frac{\lambda}{3}\right) + \frac{1}{1-\eta} P\left(\max_{1\leq i\leq n} |X_i| \geq \frac{\lambda}{3p}\right). \tag{3.3.2}$$

<u>Proof</u>. Let again $\quad E_i = \{ \max_{j<i} |S_j| < \lambda \leq |S_i|\}, \quad i = \overline{1,n}$
We can write

$$P\left(\max_{1\leq i\leq n} |S_i| \geq \lambda\right) \leq P\left(|S_n| \geq \frac{\lambda}{3}\right) + P\left(\bigcup_{i=1}^{n-1} \left(E_i \cap \{|S_n - S_i| \geq \tfrac{2\lambda}{3}\}\right)\right) =$$

$$= P\left(|S_n| \geq \frac{\lambda}{3}\right) + P\left(\bigcup_{i=1}^{n-p-1} \left(E_i \cap \{|S_n - S_i| \geq \tfrac{2\lambda}{3}\}\right)\right) +$$

$$+ P\left(\bigcup_{i=n-p}^{n} \left(E_i \cap \{|S_n - S_i| \geq \tfrac{2\lambda}{3}\}\right)\right) \leq$$

$$\leq P\left(|S_n| \geq \frac{\lambda}{3}\right) + P\left(\bigcup_{i=n-p}^{n} E_i \cap \left\{\max_{1\leq i\leq n} |X_i| \geq \tfrac{\lambda}{3p}\right\}\right) +$$

$$+ P\left(\bigcup_{i=1}^{n-p-1} \left(E_i \cap \{|S_n - S_{i+p}| \geq \tfrac{\lambda}{3}\}\right)\right) +$$

$$+ P\left(\bigcup_{i=1}^{n-p-1} \left(E_i \cap \{|S_{i+p} - S_i| \geq \tfrac{\lambda}{3}\}\right)\right) \leq P\left(|S_n| \geq \frac{\lambda}{3}\right) +$$

$$+ \sum_{i=1}^{n-p-1} P\left(E_i \cap \{|S_n - S_{i+p}| \geq \tfrac{\lambda}{3}\}\right) + P\left(\max_{1\leq i\leq n} |X_i| \geq \tfrac{\lambda}{3p}\right) \leq$$

$$\leq P\left(|S_n| \geq \frac{\lambda}{3}\right) + P\left(\max_{1\leq i\leq n} |X_i| \geq \tfrac{\lambda}{3p}\right) +$$

$$+ \sum_{i=1}^{n-p-1} P(E_i)\left(P\left(|S_n - S_{i+p}| \geq \tfrac{\lambda}{3}\right) + \varphi(P)\right) \leq$$

$$\leq P\left(|S_n| \geq \frac{\lambda}{3}\right) + P\left(\max_{1\leq i\leq n} |X_i| \geq \tfrac{\lambda}{3p}\right) +$$

$$+ P\left(\max_{1\leq i\leq n} |S_i| \geq \lambda\right)\left(\max_{1\leq i\leq n-p} P\left(|S_n - S_{i+p}| \geq \tfrac{\lambda}{3}\right) + \varphi(P)\right).$$

So the **Theorem 3.3.2** is proven.

Theorem 3.3.3. Let $X_1, \ldots, X_n$ be the random variables. Suppose that for some $\lambda > 0$ and $p \in \mathbb{N}$

$$\varphi(p) + \max_{1 \leq i \leq n} P\left(|S_n - S_i| \geq \tfrac{\lambda}{3}\right) \leq \eta < 1. \qquad (3.3.3)$$

Then

$$P(|S_n| \geq \lambda) \leq \frac{\eta}{1-\eta} P\left(|S_n| > \tfrac{\lambda}{3}\right) + \frac{1}{1-\eta} P\left(\max_{1 \leq i \leq n} |X_i| > \tfrac{\lambda}{3p}\right).$$

Proof. We have

$$P(|S_n| \geq \lambda) \leq P\left(\max_{1 \leq i \leq n} |X_i| > \tfrac{\lambda}{2p}\right) \leq$$

$$\leq P\left(|S_n| \geq \lambda, \ \max_{1 \leq i \leq n-p} |S_i| \geq \tfrac{\lambda}{2}, \ \max_{1 \leq i \leq n} |X_i| \leq \tfrac{\lambda}{2p}\right).$$

Since $\quad |S_n - S_{j+p-1}| \geq |S_n| - |S_{j-1}| - p \max_{1 \leq i \leq n} |X_i|$

for all $\quad 1 \leq j \leq n - p \qquad$ it follows

$$P(|S_n| \geq \lambda) \leq \sum_{i=1}^{n-p} P\left(E_j \cap \left\{|S_n - S_{j+p-1}| > \tfrac{\lambda}{2}\right\}\right) +$$

$$+ P\left(\max_{1 \leq i \leq n} |X_i| > \tfrac{\lambda}{2}\right).$$

By the definition of φ -mixing and (3.3.3) we have

$$P(|S_n| \geq \lambda) \leq \eta \, P\left(\max_{1 \leq i \leq n} |S_i| > \tfrac{2\lambda}{3}\right) + P\left(\max_{1 \leq i \leq n} |X_i| > \tfrac{\lambda}{2p}\right).$$

This and Theorem 3.3.2 concludes the proof.
We will need later also the following inequalities.

Theorem 3.3.4. Let $X_1, \ldots, X_n$ be the random variables and suppose that for some $\lambda > 0$, $p \in \mathbb{N}$,

$$\rho(p) + \max_{1 \leq i \leq n-1} P\left(|S_n - S_i| \geq \lambda\right) \leq \eta < 1.$$

Then

$$P\left(\max_{1 \leq j \leq n} |S_j| \geq 3\lambda\right) \leq \frac{1}{1-\eta} P(|S_n| \geq \lambda) +$$

$$+ \frac{\rho(P)}{1-\eta} \sum_{i=0}^{k-2} P(|S_n - S_{(i+2)p}| \geqslant \lambda) + \qquad\qquad (3.3.3)'$$

$$+ \frac{1}{1-\eta} \cdot \frac{n}{p} P\left(\sum_{i=1}^{P} |X_i| \geqslant \lambda \right), \qquad k = \left[\frac{n}{p}\right].$$

<u>Proof.</u>   Let $E_i = \left\{ \max_{1 \leqslant j \leqslant i-1} |S_j| < 3\lambda \leqslant S_i \right\}, \quad S_i = \sum_{j=1}^{i} X_j$.
We have

$$P\left(\max_{1 \leqslant j \leqslant n} |S_j| \geqslant 3\lambda \right) \leqslant P(|S_n| \geqslant \lambda) +$$

$$+ \sum_{i=0}^{k-2} P\left(\left\{ \bigcup_{j=1}^{P} E_{ip+j} \right\} \cap \left\{ |S_n - S_{(i+2)p}| \geqslant \lambda \right\} \right) +$$

$$+ \sum_{j=(k-1)p+1}^{n} P\left(E_j \cap \left\{ |S_n - S_j| \geqslant 2\lambda \right\} \right).$$

$$\sum_{i=0}^{k-2} P\left(\bigcup_{j=1}^{P} E_{ip+j} \cap \left\{ |S_{(i+2)p} - S_{ip+j}| \geqslant \lambda \right\} \right).$$

Using the definition of ρ -mixing coefficient we can write

$$P\left(\max_{1 \leqslant j \leqslant n} |S_j| \geqslant 3\lambda \right) \leqslant P(|S_n| \geqslant \lambda) +$$

$$+ \max_{2p \leqslant i \leqslant n-1} P\left(\left\{ |S_n - S_i| \geqslant \lambda \right\} \right) \sum_{i=1}^{n} P(E_i) +$$

$$+ \rho(P) \sum_{i=0}^{k-2} P^{\frac{1}{2}}\left(\bigcup_{j=1}^{P} E_{ip+j} \right) P^{\frac{1}{2}}\left(\left\{ |S_n - S_{(i+2)p}| \geqslant \lambda \right\} \right) +$$

$$+ \sum_{i=0}^{k-2} P\left(\left\{ \max_{1 \leqslant j \leqslant P} |S_{(i+2)p} - S_{ip+j}| \geqslant \lambda \right) + \right.$$

$$+ P\left(\left\{ \max_{(k-1)p+1 \leqslant j \leqslant n} |S_n - S_i| > \lambda \right\} \right).$$

Since

$$P^{\frac{1}{2}}\left(\bigcup_{j=1}^{p} E_{ip+j}\right) P^{\frac{1}{2}}\left(|S_n - S_{(i+2)p}| \geq \lambda\right) \leq$$

$$\leq P\left(\bigcup_{j=1}^{p} E_{ip+j}\right) + P\left(|S_n - S_{(i+2)p}| \geq \lambda\right)$$

we find

$$P\left(\max_{1 \leq j \leq n} |S_j| \geq 3\lambda\right) \leq P\left(|S_n| \geq \lambda\right) +$$

$$+ P\left(\max_{1 \leq j \leq n} |S_j| \geq 3\lambda\right)\left(\max_{1 \leq i \leq n-1} P(|S_n - S_i| \geq \lambda) + \right.$$

$$+ \rho(p) + \rho(p) \sum_{i=0}^{k-2} P(|S_n - S_{(i+2)p}| \geq \lambda) +$$

$$\left[\frac{n}{p}\right] P\left(\sum_{i=1}^{p} |X_i| \geq \lambda\right).$$

By this the proof is finished.

Note the similar inequality can be found in /99/ .

<u>Lemma 3.3.1.</u> Let $(X_1, X_2, \ldots, X_n)$ be a random vector such that $E\left|\prod_{j=1}^{k} X_j\right| < \infty$, $k = 1, \ldots, n$. Then

$$|EX_1 X_2 \cdots X_n - EX_1 EX_2 \cdots EX_n| \leq |EX_1 X_2 \cdots X_n - EX_1 EX_2 \cdots EX_n| +$$

$$+ \sum_{j=1}^{n-2} \prod_{k=1}^{j} |EX_k| |EX_{j+1} \cdots X_n - EX_{j+1} EX_{j+2} \cdots X_n|. \quad (3.3.4)$$

<u>Proof.</u> Let

$$T_k = EX_k \cdots X_n - EX_k EX_{k+1} \cdots EX_n ,$$

$$G_k = EX_k \cdots X_n - EX_k EX_{k+1} \cdots X_n .$$

Then

$$T_k = G_k + EX_k T_{k+1} \; , \qquad k = 1, 2, \ldots, n-1.$$

Hence

$$T_1 = G_1 + \sum_{j=1}^{n-2} \prod_{k=1}^{j} EX_k G_{j+1}$$

and inequality (3.3.4) holds.

Note, if under conditions of Lemma 3.3.1 is supposed additionally that $|EX_k| \leqslant 1$, $k = \overline{1, n}$, then from (3.3.4) we obtain

$$|EX_1 \ldots X_n - EX_1 \ldots EX_n| \leqslant \sum_{j=1}^{n-1} |EX_j \ldots X_n - EX_j EX_{j+1} \ldots X_n|. \qquad (3.3.5)$$

<u>Lemma 3.3.2.</u> Let the random vector $(X_1, \ldots, X_n)$ be such that $E\left|\prod_{j=1}^{k} X_j\right| < \infty$, $k = \overline{1, n}$. Then

$$|EX_1 X_2 \ldots X_n - EX_1 EX_2 \ldots EX_n| \leqslant \sum_{j=2}^{n} \Big| E(X_1-1)(X_j-1)\prod_{s=j+1}^{n} X_s -$$

$$- E(X_1-1)E(X_j-1)\prod_{s=j+1}^{n} X_s \Big| +$$

$$+ \sum_{j=1}^{n-2} \prod_{k=1}^{j} |EX_k| \sum_{s=j+2}^{n} \Big| E(X_{j+1}-1)(X_s-1)\prod_{i=s+1}^{n} X_s -$$

$$- E(X_{j+1}-1)E(X_s-1)\prod_{i=s+1}^{n} X_s \Big|.$$

<u>Proof.</u> We have

$$G_k = EX_k \ldots X_n - EX_k EX_{k+1} \ldots X_n = E(X_k-1)X_{k+1} \ldots X_n -$$

$$-E(X_k-1)EX_{k+1} \ldots X_n = \sum_{j=k+1}^{n} \Big(E(X_k-1)(X_j-1)\prod_{s=j+1}^{n} X_s -$$

$$- E(X_k-1)E(X_j-1)\prod_{s=j+1}^{n} X_s \Big).$$

Hence

$$|G_k| \leq \sum_{j=k+1}^{n} |E(X_k-1)(X_j-1)\prod_{s=j+1}^{n} X_s -$$ (3.3.6)

$$E(X_1-1)E(X_j-1)\prod_{s=j+1}^{n} X_s|.$$

Substituting (3.3.6) into (3.3.4) we obtain the required inequality.

If we suppose that $|EX_k| \leq 1$, $k = 1, 2, \ldots, n$, then under the condition of Lemma 3.3.2

$$|EX_1 X_2 \ldots X_n - EX_1 EX_2 \ldots EX_n| \leq$$ (3.3.7)

$$\leq \sum_{j=1}^{n-1} \sum_{s=j+1}^{n} |E(X_j-1)(X_s-1)\prod_{p=s+1}^{n} X_p - E(X_j-1)E(X_s-1)\prod_{p=s+1}^{n} X_p|.$$

Finally we formulate some results of Yokoyama /131/.

Theorem 3.3.5. Let $\xi_t, t \in \mathbb{Z}^1$, be a strictly stationary strong mixing sequence with $E\xi_0 = 0$ and $E|\xi_0|^{v+\delta} < \infty$ for some $v > 2$ and $\delta > 0$. If

$$\sum_{i=0}^{\infty} (i+1)^{v/2-1} [\alpha(j)]^{\delta/2+\delta} < \infty$$

then there exists a constant K such that

$$E|S_n|^v \leq Kn^{v/2}, \quad n \geq 1.$$ (3.3.8)

Theorem 3.3.6. Let $\xi_t, t \in \mathbb{Z}^1$, be a strictly stationary strong mixing sequence with $E\xi_0 = 0$ and $|\xi_0| \leq C < \infty$ a.s. If

$$\sum_{i=0}^{\infty} (i+1)^{v/2-1} \alpha(j) < \infty$$

then (3.3.8) holds.

Note also that some interesting inequalities are presented in /140/ and /141/.

<u>4. METHODS</u>

At present the methods for proving the limit theorems for dependent random variables can be subdivided into two types. The methods of the first type use the approximation with random variables, the asymptotic behaviour of which being known, for instance, the approximation with independent random variables (Bernstein's method) or with ergodic martingale-differences (Gordin's method).The methods of the second type use the direct approximation (Stein's method,the classical methods of moments). In this book we are interested in a number of problems connected with c.l.th. and therefore we will demonstrate the methods using this example, though the applications of these methods are not exhausted by it.

<u>4.1 Bernstein's Method</u>

Now we will formulate one simple lemma in which the main points of c.l.th. proof by Bernstein's method are reflected. Let
ξ_t, $t \in \mathbb{Z}^\nu$, be a random field, $[-n, n]^\nu \subset \mathbb{Z}^\nu$ be the ν -dimensional cube with the side 2nd and the center in the origin and let

$$S_I = \sum_{t \in I} \xi_t \;, \quad I \subset \mathbb{Z}^\nu, \; |I| < \infty \;,$$

$$\sigma_{n,\nu}^2 = DS_{[-n,n]^\nu} \;, \quad n \in \mathbb{N} \;.$$

<u>Lemma 4.1.1.</u>   Suppose that ξ_t, $t \in \mathbb{Z}^\nu$, is a centred random field and for some $\delta \geqslant 0$, $E|\xi_t|^{2+\delta} < \infty$, $t \in \mathbb{Z}^\nu$. Suppose in addition there exists a sequence of subsets $\Delta_j^{(n)}$ of ν -dimensional cube $[-n, n]^\nu$, $\Delta_j^{(n)} \subset [-n, n]^\nu$, $j = 1, 2, \ldots, k$, $k = k(n)$, $k(n) \to \infty$ as $n \to \infty$ such that the following conditions are valid:

1) $\quad \sigma_{n,\nu}^{-2} DS_{(I_n^\nu)'} \longrightarrow 0 \qquad$ as $\quad n \to \infty$,

where $I_n^\nu = \bigcup_{j=1}^{k} \Delta_j^{(n)}$ and $(I_n^\nu)'$ is the complement of I_n^ν with respect to $[-n, n]^\nu$;

2) for any fixed $t \in R^1$

$$\mathfrak{M}_k^{(t)} = \left| E \prod_{j=1}^{k} e^{itS(n,j)} - \prod_{j=1}^{k} E\, e^{itS(n,j)} \right| \to 0 , \qquad \text{as } n \to \infty,$$

here $\quad S(n,j) = \sigma_{n,\nu}^{-1}\, S\,\Delta_j^{(n)} , \quad j = \overline{1,k} ;$

3) for any $\varepsilon > 0$

$$L_k(\varepsilon) = \sum_{j=1}^{k} \int\limits_{|S(n,j)| \geq \varepsilon} |S(n,j)|^{2+\delta}\, P(d\omega) \to 0 , \qquad \text{when } n \to \infty.$$

Then for every $x \in R^1$

$$\lim_{n \to \infty} P(\sigma_{n,\nu}^{-1}\, S_{[-n,n]\nu} < x) = \frac{1}{\sqrt{2\pi}} \int\limits_{-\infty}^{x} e^{-u^2/2}\, du, \qquad (4.1.1)$$

i.e., the c.l.th. holds.

<u>Proof</u>. It is easy to verify the validity of this lemma. We can write

$$\frac{1}{\sigma_{n,\nu}}\, S_{[-n,n]\nu} = \frac{1}{\sigma_{n,\nu}}\, S_{I_n^\nu} + \frac{1}{\sigma_{n,\nu}}\, S_{(I_n^\nu)'} . \qquad (4.1.2)$$

By condition 1) of this lemma if follows that the second summand in (4.1.2) tends to zero in probability when $n \to \infty$ and thus the limit behaviour of $\sigma_{n,\nu}^{-1}\, S_{[-n,n]\nu}$ coincides with that of random variables $\sigma_{n,\nu}^{-1}\, S_{I_n^\nu}$. The latter as it follows from the conditions 3) and 4) is asymptotically normal because its limit coincides with the limit of independent random variables for which the Lindeberg condition is fulfilled.

Note, that Lemma 4.1.1 is usually specified by a special choice of subsets $\Delta_j^{(n)}$, $j = \overline{1,k}$. The following choice is mainly used. Let $p = p(n)$, $q = q(n)$, $n \in \mathbb{N}$, be some function assuming natural values and such ones that

$$p,\ q \to \infty, \qquad p = o(n), \qquad q = o(p) \qquad \text{as } n \to \infty.$$

Further such a pair of functions will be called a standard one. Let

$$\hat{k} = \hat{k}(n) = \left[\frac{2n}{p+q} \right] ,$$

where $[\cdot]$ denotes the integer part of some number,

$$I_n(i) = [-n + ip + iq, \ -n + (i+1)p + iq] \subset \mathbb{Z}^1;$$

$$i = 0, \ldots, \hat{k}-1 \ ; \qquad I_n = \bigcup_{i=0}^{\hat{k}-1} I_n(i).$$

Let I_n^ν be the Cartesian product of ν-copies of I_n. Then I_n^ν is the union of $(\hat{k}-1)^\nu$ ν-dimensional cubes with side $p = p(n)$. Let these cubes be somehow enumerated (for $\nu = 1$ enumeration is natural). Further $\Delta_j^{(n)}$ always denotes a cube with side $p = p(n)$ and number j, $j = \overline{1,k}$, $k = k(n) = (\hat{k}(n)-1)^\nu$. It is evident that

$$I_n^\nu = \bigcup_{j=1}^{k} \Delta_j^{(n)}.$$

For such a concrete definition the ν-dimensional cube $[-n,n]^\nu$ turns out to be divided into "rooms" and "corridors", the "corridors" being non-essential part and according to condition 1) of Lemma 4.1.1 the limit distribution being determined by "rooms". Note that condition 1) of Lemma 4.1.1 directly comes from linear behaviour of variance which as rule is assumed (or easily comes from conditions) for the c.l.th. proof. Thus the main difficulty of c.l.th. proof for random fields by Bernstein's method is to check the conditions 2), 3) of Lemma 4.1.1. In doing this for mixing random fields and rapidly increasing function $p = p(n)$ the verification of condition 2) proves to be simple and the difficult part refers to the verification of Lindeberg condition, i.e., condition 3) of Lemma 4.1.1. In traditional c.l.th. proofs by Bernstein's method the rapidly increasing functions are usually used. In the next chapter we will demonstrate that the usage of slowly increasing functions $p(n)$ is no less effective, the present situation turning out quite opposite: the Lindeberg condition is checked trivially and the main difficulty lies in the verification of condition 2) of Lemma 4.1.1. In connection with this we will give some preliminary lemmas of general character. For implicity we will consider stationary random fields.

Lemma 4.1.2. Let ξ_t, $t \in \mathbb{Z}^\nu$, be a s.r.f. such that

$$E|\xi_0|^{2+\delta} < \infty, \ \delta > 0, \quad DS_I \sim \sigma_0^2 |I|, \quad |I| \to \infty, \ \sigma_0^2 > 0.$$

If for some standard pair of functions p and q the following two conditions are fulfilled:

1) $\mathfrak{M}_k^{(t)} \to 0$ as $n \to \infty$ for fixed $t \in R^1$,

2) for any $\varepsilon > 0$, $L_k(\varepsilon) \to 0$ as $n \to \infty$,

then sequence $\sigma_{n,\nu}^{-1} S_{[-n,n]\nu}$ is asymptotically normal.
Inversely, if sequence $\sigma_{n,\nu}^{-1} S_{[-n,n]\nu}$ is asymptotically normal, then for any standard pair of functions p,q, the relations 1) and 2) are fulfilled.

<u>Proof.</u> The first statement is a direct consequence of Lemma 4.1.1 . Let us prove the second statement. Suppose that $\sigma_{n,\nu}^{-1} S_{[-n,n]\nu}$ is asymptotically normal and let p, q be some standard pair of functions. With the help of the usual procedure we can write

$$S_{[-n,n]\nu} = \sum_{j=1}^{k} S_{\Delta_j^{(n)}} + S_{(\bigcup_{j=1}^{k} \Delta_j^{(n)})'} .$$

Further it is obvious that for the proof it is sufficient to establish the relation 2) as according to our assumption 2) implies 1). In its turn it is sufficient to prove 2) for $\delta = 0$. Note that $\sigma_{n,\nu}^{-1} S_{[-p,p]\nu}$ is the subsequence of $\sigma_{n,\nu}^{-1} S_{[-n,n]\nu}$ and hence it weakly converges to N (o,1). Thus it is possible to write

$$\lim_{n \to \infty} D \frac{S_{[-p,p]\nu}}{\sigma_{p,\nu}} = D \lim_{n \to \infty} \frac{S_{[-p,p]\nu}}{\sigma_{p,\nu}} = 1 .$$

Whence the uniform integrability of sequence $\sigma_{p,\nu}^{-1} S_{[-p,p]\nu}$ holds (see sect.1.7). By virtue of stationary state of r.f. for the verification of condition 2) it is sufficient to show that

$$\lim_{n \to \infty} \sigma_{p,\nu}^{-2} E S_{[-p,p]\nu}^2 \, I\left(|S_{[-p,p]\nu}| \geq \varepsilon \sigma_{p,\nu}\right) = 0 . \tag{4.1.3}$$

We have

$$\lim_{n \to \infty} \sigma_{p,\nu}^{-2} E S_{[-p,p]\nu}^2 I\left(|S_{[-p,p]\nu}| \geq \varepsilon \sigma_{p,\nu}\right) \leq$$

$$\lim_{n \to \infty} \sigma_{p,\nu}^{-2} E S_{[-p,p]\nu}^2 I\left(|S_{[-p,p]\nu}| \geq K\right) \leq$$

$$\leq \int_{|\xi| \geq K} \xi^2 P(d\omega), \qquad \xi \sim N(0,1) .$$

Hence because of the arbitrariness of K the validity of this lemma follows.

$\underline{\text{Lemma 4.1.3.}}$ Let ξ_t, $t \in \mathbb{Z}^\nu$ be a s.r.f. such that $E\xi_0 = 0$, $E|\xi_0|^{2+\delta} < \infty$, $\delta \geq 0$, $DS_I \sim \sigma_0^2 |I|$, $|I| \to \infty$, $\sigma_0^2 > 0$.
Then for

$$\frac{S_{[-n,n]^\nu}}{\sigma_{n,\nu}} \xrightarrow[n \to \infty]{\mathcal{D}} N(0,1)$$

it is necessary and sufficient to fulfil the following relation $\mathfrak{m}_k^{(t)} \to 0$, as $n \to \infty$ and fixed $t \in R^1$ for any standard pair of functions p and q.

$\underline{\text{Proof.}}$ **The necessity of the** condition is the consequence of **the above** mentioned lemma. The sufficiency is also evident since there exists the function $p = p(n)$, $p \to \infty$, $p = o(n)$, $n \to \infty$ such that the condition 2) of Lemma 4.1.2 is valid. Taking now as $q = q(n)$ any function such that $q \to \infty$, $q = o(p)$, $n \to \infty$, we get a standard pair for which the conditions of previous lemma are fulfilled.

Note that if $\delta > 0$ it is possible to indicate the rate of convergence to infinity of the function $p(n)$. Really, we have

$$\sigma_{p,\nu}^{-2} E\left(S_{[-p,p]^\nu}\right)^2 I\left(|S_{[-p,p]^\nu}| \geq \varepsilon \sigma_{[-n,n]^\nu}\right) \leq$$

$$\leq C_1 \left(n^{\delta/2} p\right)^\nu E|S_{[-p,p]^\nu}|^{2+\delta}.$$

By the Minkowski inequality we obtain

$$\left(n^{\delta/2} p\right)^{-\nu} E|S_{[-p,p]^\nu}|^{2+\delta} \leq C_2 \left(n^{\delta/2} p\right)^{-\nu} \left(p^{2+\delta}\right)^\nu = C_2 \left(\frac{p^{1+\delta}}{n^{\delta/2}}\right)^\nu.$$

The latter tends to zero as $n \to \infty$ if $p = o(n^{\delta/2 + \delta})$. In case, the components of r.f. are bounded with probability 1 we may put $p = o(n^{1/2})$. This fact was first mentioned by Bulinskii and Zurbenko (/19/). Note that in all statements of this paragraph no suppositions of weak dependence of random field components were made.

4.2 Gordin's Method

As it has already been noted Gordin's method is based on the approximation of considered random variables by martingale-differences. Hence there is some narrowness of the range of its usage. For instance, it does not work in multi-dimensional case ($\nu > 1$). The sequence of random variables ξ_t, $t \in \mathbb{Z}^1$, is called martingale-difference if $E\xi_t = 0$, $t \in \mathbb{Z}^1$ and

$$E|\xi_t| < \infty, \quad t \in \mathbb{Z}^1,$$
$$E(\xi_t / \xi_{t-1}, \xi_{t-2}, \dots) = \xi_t, \quad t \in \mathbb{Z}^1. \qquad (4.2.1)$$

Below we give a simple proof of **the well-known Billingsley-Ibragimov theorem** (/6/,/64/) concerning the asymptotical normality of such sequences, which is the basis of Gordin's method.

<u>Theorem 4.2.1</u> (Billingsley-Ibragimov). Let ξ_t, $t \in \mathbb{Z}^1$, be a stationary, ergodic sequence of martingale-differences and $E\xi_t = 0$, $E\xi_t^2 = 1$, $t \in \mathbb{Z}^1$. Then

$$P(\sigma_n^{-1} S_n < x) \to N(0,1), \quad n \to \infty.$$

<u>Proof.</u> Let

$$z_s^{(n)} = E \prod_{j=1}^{s} e^{it \frac{\xi_j}{\sqrt{n}}} - \prod_{j=1}^{s} E e^{it \frac{\xi_j}{\sqrt{n}}}, \quad s = 1, \dots, n.$$

To prove this theorem it is sufficient to show that

$$z_n^{(n)} \to 0 \quad , \text{ as } \quad n \to \infty. \qquad (4.2.2)$$

Since

$$E\left(\xi_s \prod_{j=1}^{s-1} e^{it \frac{\xi_i}{\sqrt{n}}}\right) = E\left(\prod_{j=1}^{s-1} e^{it \frac{\xi_i}{\sqrt{n}}} E\left(\xi_s / \xi_{s-1}, \dots\right)\right) = 0,$$

then we can write

$$z_s^{(n)} = E\left(e^{it \frac{\xi_s}{\sqrt{n}}} - 1\right) \prod_{j=1}^{s-1} e^{it \frac{\xi_j}{\sqrt{n}}} - E\left(e^{it \frac{\xi_s}{\sqrt{n}}} - 1\right) \prod_{j=1}^{s-1} E e^{it \frac{\xi_j}{\sqrt{n}}} +$$

$$+ z_{s-1}^{(n)} = E\left(e^{it \frac{\xi_s}{\sqrt{n}}} - 1 - it \frac{\xi_s}{\sqrt{n}} + \frac{t^2}{2n} \xi_s^2\right) \prod_{j=1}^{s-1} e^{it \frac{\xi_j}{\sqrt{n}}} -$$

$$- E\left(e^{it \frac{\xi_s}{\sqrt{n}}} - 1 - it \frac{\xi_s}{\sqrt{n}} + \frac{t}{2n} \xi_s^2\right) \prod_{j=1}^{s-1} E e^{it \frac{\xi_j}{\sqrt{n}}} -$$

$$-\frac{t}{2n}\,E\,\xi_s^2\prod_{j=1}^{s-1}e^{it\frac{\xi_j}{\sqrt{n}}}+\frac{t}{2n}\,E\,\xi_s^2\prod_{j=1}^{s-1}E\,e^{it\frac{\xi_j}{\sqrt{n}}}+z_{s-1}^{(n)}\,.$$

Denote

$$\alpha_{n,s}=E\left(e^{it\frac{\xi_s}{\sqrt{n}}}-1-it\frac{\xi_s}{\sqrt{n}}+\frac{t^2}{2n}\xi_s^2\right)\prod_{j=1}^{s-1}e^{it\frac{\xi_j}{\sqrt{n}}}-$$

$$-E\left(e^{it\frac{\xi_s}{\sqrt{n}}}-1-it\frac{\xi_s}{\sqrt{n}}+\frac{t^2}{2n}\xi_s^2\right)\prod_{j=1}^{s-1}E\,e^{it\frac{\xi_j}{\sqrt{n}}}\,,$$

$$\beta_{n,s}=-\frac{t^2}{2n}\left(E\,\xi_s^2\prod_{j=1}^{s-1}e^{it\frac{\xi_j}{\sqrt{n}}}-E\,\xi_s^2\,E\prod_{j=1}^{s-1}e^{it\frac{\xi_j}{\sqrt{n}}}\right).$$

Then we have the following recurrent relation:

$$z_s^{(n)}=\alpha_{n,s}+\beta_{n,s}+z_{s-1}^{(n)}\left(1-\frac{t^2}{2n}\right).$$

From here we obtain

$$z_n^{(n)}=\sum_{s=1}^{n}(\alpha_{n,s}+\beta_{n,s})\left(1-\frac{t^2}{2n}\right)^{n-s}$$

and

$$\left|z_n^{(n)}\right|\le\sum_{s=1}^{n}|\alpha_{n,s}|+\left|\sum_{s=1}^{n}\beta_{n,s}\right|.\qquad(4.2.3)$$

Taking advantage of the known inequality

$$\left|\exp\{ix\}-1-ix-\cdots-\frac{(ix)^{n-1}}{(n-1)!}\right|\le\frac{|x|^n}{n!}\qquad(4.2.4)$$

we can write

$$\sum_{s=1}^{n}|\alpha_{n,s}|\le2\sum_{s=1}^{n}\left(\frac{t^2}{2}E|\xi_0|^2I(|\xi_0|\ge\varepsilon\sqrt{n})+\frac{\varepsilon t^3}{n}\right)\le$$

$$(4.2.5)$$

$$\le2t^2E|\xi_0|^2I(|\xi_0|>\varepsilon\sqrt{n})+2\varepsilon t^3.$$

Further

$$\sum_{s=1}^{n}\beta_{n,s}=-\frac{t^2}{2n}\sum_{s=1}^{n}\left(E\,\xi_s^2\,I(|\xi_s|>N)\prod_{j=1}^{s-1}e^{it\frac{\xi_j}{\sqrt{n}}}-\right.$$

$$-E\xi_s^2 I(|\xi_s|>N)E\prod_{j=1}^{s-1}e^{it\frac{\xi_j}{\sqrt{n}}})-$$

$$-\frac{t^2}{2n}\sum_{s=1}^{n}(E\xi_s^2 I(|\xi_s|\leq N)\prod_{j=1}^{s-1}e^{it\frac{\xi_j}{\sqrt{n}}}-E\xi_s^2 I(|\xi_s|\leq N)E\prod_{j=1}^{s-1}e^{it\frac{\xi_j}{\sqrt{n}}}).$$

Hence

$$\left|\sum_{s=1}^{n}\beta_{n,s}\right|\leq t^2 E|\xi_0|^2 I(|\xi_0|>N)+$$

$$+\frac{t^2}{2n}\left|\sum_{s=1}^{n}E\prod_{j=1}^{s-1}e^{it\frac{\xi_j}{\sqrt{n}}}(E(\xi_s^2 I(|\xi_s|\leq N)/\xi_{s-1},\ldots)-\right.$$

$$\left.-E\xi_s^2 I(|\xi_s|\leq N)\right|. \tag{4.2.6}$$

Denote

$$\eta_s = \xi_s^2 I(|\xi_s|\leq N), \quad E_s\eta_s = E\left(\eta_s / \xi_{s-1},\ldots\right), \quad s=1,2,\ldots$$

Using Abel's transformation
$$\sum_{j=1}^{m}a_j b_j = a_m B_m - \sum_{j=1}^{m-1}(a_{j+1}-a_j)B_j, \quad B_j = \sum_{s=1}^{j}b_s, \quad a_j, b_j\in R^1$$
in the second member of inequality (4.2.6) we obtain

$$\left|\sum_{s=1}^{n}\beta_{n,s}\right|\leq t^2 E|\xi_0|^2 I(|\xi_0|>N)+$$

$$+\frac{t^2}{2}\left|E\prod_{j=1}^{n-1}e^{it\frac{\xi_j}{\sqrt{n}}}\sum_{s=1}^{n}(E_s\eta_s-E\eta_s)\right|+$$

$$+\frac{t^2}{2}\left|E\sum_{k=1}^{K}\left(\prod_{j=1}^{K}e^{it\frac{\xi_j}{\sqrt{n}}}-\prod_{j=1}^{K-1}e^{it\frac{\xi_j}{\sqrt{n}}}\right)\sum_{s=1}^{K}(E_s\eta_s-E\eta_s)\right|\leq$$

$$\leq t^2 E|\xi_0|^2 I(|\xi_0|>N)+\frac{t^2}{2}E\left|\sum_{s=1}^{n}(E_s\eta_s-E\eta_s)\right|+$$

$$+\frac{t^2}{2n}\left|E\sum_{k=1}^{n-1}\prod_{j=1}^{k-1}e^{it\frac{\xi_j}{\sqrt{n}}}\left(e^{it\frac{\xi_k}{\sqrt{n}}}-1-\frac{it\xi_k}{\sqrt{n}}\right)\sum_{s=1}^{k-1}(E_s\eta_s-E\eta_s)\right|+$$

$$+ \frac{t^2}{2n} \left| E \sum_{k=1}^{n-1} \prod_{j=1}^{k-1} e^{it\frac{\xi_j}{\sqrt{n}}} \left(e^{it\frac{\xi_k}{\sqrt{n}}}-1\right)\left(E_k \eta_k - E \eta_k\right)\right|.$$

Applying Jensen's inequality and (4.2.4) **we have**

$$\left| \sum_{s=1}^{n} \beta_{n,s} \right| \leqslant t^2 E |\xi_0|^2 I(|\xi_0| > N) + \frac{t^2}{2} E \left| \frac{\sum_{s=1}^{n}(\eta_s - E\eta_s)}{n} \right| +$$

$$+ \frac{t^2}{2n} \sum_{k=1}^{n-1} E \left| \frac{\sum_{s=1}^{k-1}(\eta_s - E\eta_s)}{k} \right| + \frac{Nt^2}{\sqrt{n}}.$$

Finally

$$\left| z_n^{(n)} \right| \leqslant C_1 t^2 \left(E |\xi_0|^2 I(|\xi_0| \geqslant \varepsilon \sqrt{n}) + \varepsilon t + \frac{N}{n} +$$

$$+ E |\xi_0|^2 I(|\xi_0| > N) + n^{-1} E \left| \sum_{s=1}^{n}\left(\xi_s^2 I(\xi_s \leqslant N) -\right.\right.$$

$$\left.\left. - E \xi_s^2 I(\xi_s \leqslant N)\right)\right|.$$

Now it is evident that (4.2.2) follows from the integral ergodic theorem.

Let $(\Omega, \mathcal{F}, P)$ be a probability space and T be an ergodic one to one measure preserving transformation. Denote by Q the linear space of elements $q \in L_2(\Omega)$ and $q \in H(\mathcal{H}_{-\infty}^k) \ominus H(\mathcal{H}_{-\infty}^j)$ for some finite k and j, $(k < j)$. The Gordin's theorem given below opens up the possibility to approximate the sums of random variables by a naturally related martingale.

<u>Theorem 4.2.2</u> (/47/). Let $f \in L_2(\Omega)$ and

$$\inf_{q \in Q} \overline{\lim_{n \to \infty}} \ n^{-1} E \left(\sum_{k=0}^{n-1} U^k (f - q) \right)^2 = 0. \tag{4.2.7}$$

Then

$$\lim_{n \to \infty} n^{-1} E \left(\sum_{k=0}^{n-1} U^k f \right)^2 = \sigma^2, \quad 0 \leqslant \sigma^2 < \infty$$

and

$$n^{-\frac{1}{2}} \sum_{k=0}^{n-1} U^k f \xrightarrow[n \to \infty]{\mathcal{D}} N(0,1).$$

Suppose that there will be found a random variable $Y_0 \in L_2(\Omega)$ such that the collection $Y_k = U^k Y_0$, $k \in \mathbb{Z}^1$, composes a stationary ergodic sequence of martingale-differences. Represent the sum $S_n = \sum_{t=1}^{n} \xi_t$ in the following way:

$$S_n = \sum_{k=1}^{n} Y_k + \sum_{k=1}^{n} U^k (\xi_0 - Y_0).$$

Now it is obvious if in probability

$$\frac{1}{\sqrt{n}} \left| \sum_{k=1}^{n} U^k (\xi_0 - Y_0) \right| \to 0, \quad n \to \infty$$

then S_n with appropriate norming is asymptotically normal. For example, it is sufficient to show

$$n^{-1} E \left(\sum_{k=1}^{n} U^k (\xi_0 - Y_0) \right)^2 \to 0, \quad n \to \infty.$$

The following theorem gives easily checked conditions of the asymptotical negligibility of the sequence

$$\frac{1}{\sqrt{n}} \sum_{k=1}^{n} U^k (\xi_0 - Y_0).$$

<u>Theorem 4.2.3</u> (/60/). Let ξ_t, $t \in \mathbb{Z}^1$, be a stationary ergodic r.p. such that $E\xi_0 = 0$, $E\xi_0^2 < \infty$. If in addition to this the following conditions are fulfilled:

1) $\sum_{k=1}^{\infty} E(\xi_k E(\xi_t / \mathcal{O}l_{-\infty}^0))$ converges for any $t \geq 0$

2) $\lim_{t \to \infty} \sum_{k=N}^{\infty} E(\xi_k E(\xi_t / \mathcal{O}l_{-\infty}^0)) = 0$,   uniformly in N

then

$$\lim_{n \to \infty} n^{-1} E S_n^2 = \sigma^2, \quad 0 \leq \sigma^2 \leq \infty,$$

and if $\sigma^2 > 0$, then the sequence $(\sigma \sqrt{n})^{-1} S_n$ is asymptotically normal.

<u>Proof.</u> From condition 2) of this theorem it follows

$$E(\xi_N E(\xi_N / \mathcal{O}l_{-\infty}^0)) = E(E(\xi_0 / \mathcal{O}l_{-\infty}^{-N}))^2 \to 0, \quad N \to \infty.$$

Thus for any $\varepsilon > 0$ we can choose N, so that

$$E(E(\xi_0 / \mathcal{O}l_{-\infty}^{-N}))^2 < \varepsilon.$$

Put

$$f = \xi_0, \quad g = \xi_0 - E(\xi_0 / \mathcal{O}l_{-\infty}^{-N}).$$

Since ξ_0 is measurable with respect to $\mathcal{U}_{-\infty}^{0}$, then

$$g \in H_0 \ominus H_{-N} , \qquad g \in Q.$$

Now we can write

$$n^{-1} E \Big(\sum_{k=0}^{n-1} U^K (\xi_0 - g) \Big)^2 = n^{-1} E \Big(\sum_{k=0}^{n-1} E (\xi_K / \mathcal{U}_{-\infty}^{-N+k}) \Big)^2 =$$

$$= E \big(E (\xi_0 / \mathcal{U}_{-\infty}^{-N}) \big)^2 + 2n^{-1} \sum_{j<k} E \big(E (\xi_j / \mathcal{U}_{-\infty}^{-N+j}) E (\xi_K / \mathcal{U}_{-\infty}^{-N+k}) \big).$$

Here the first member is arbitrarily small. Let us estimate the second one:

$$2n^{-1} \sum_{j<k} E \big(E (\xi_0 / \mathcal{U}_{-\infty}^{-N}) \big) E (\xi_{K-j} / \mathcal{U}_{-\infty}^{-N+k-j}) =$$

$$= 2n^{-1} \sum_{k=1}^{n-1} (n-k) E \big(E (\xi_0 / \mathcal{U}_{-\infty}^{-N}) E (\xi_K / \mathcal{U}_{-\infty}^{-N+k}) \big) =$$

$$= 2n^{-1} \sum_{k=1}^{n-1} (n-k) E \big(\xi_K E (\xi_0 / \mathcal{U}_{-\infty}^{-N}) \big) =$$

$$= 2n^{-1} \sum_{k=1}^{n-1} (n-k) E \big(\xi_{k+N} E (\xi_N / \mathcal{U}_{-\infty}^{0}) \big) =$$

$$= 2 \sum_{k=N+1}^{N+n-1} E \big(\xi_K E (\xi_N / \mathcal{U}_{-\infty}^{0}) \big) - 2n^{-1} \sum_{k=1}^{n-1} k E \big(\xi_{k+n} E (\xi_N / \mathcal{U}_{-\infty}^{0}) \big).$$

The first term in the last relation tends to zero when $n \to \infty$.
As to the second member it tends to zero according to condition
1), Kronecker's lemma and the fact that

$$\sum_{k=1}^{\infty} E \big(\xi_{k+N} E (\xi_N / \mathcal{U}_{-\infty}^{0}) \big)$$

differs only in finite number of summands from

$$\sum_{k=1}^{\infty} E \big(\xi_K E (\xi_N / \mathcal{U}_{-\infty}^{0}) \big).$$

Hence taking sufficiently large N we obtain the conclusion
of Theorem 4.2.2.

In the next chapter some limit theorems will be presented, in
the proof of which this approach is used. For the applications
of martingale type techniques to limit theorems see also/39/.

4.3 Stein's Method

This method was first suggested by Stein in /120/ with the pur-
pose to estimate the rate of convergence in the c.l.th. for dependent random variables. Nevertheless Stein's method turned out
quite useful to establish the various cases of the c.l.th. for
random fields with some mixing conditions (see /9/ , /122/). The
main idea of Stein's method is well illustrated by the following
lemma (/9/).

Lemma 4.3.1. Let μ_n, $n \in \mathbb{N}$, be the sequence of probability measures on R^1 and

1) $\displaystyle \sup_n \int_{-\infty}^{\infty} |x|^2 \mu_n(dx) < \infty,$

2) $\displaystyle \lim_{n \to \infty} \int_{-\infty}^{\infty} (i\lambda - x) e^{i\lambda x} \mu_n(dx) = 0$ for all $\lambda \in R^1.$

Then the sequence μ_n, $n = 1, 2, \ldots$, weakly converges to
standard normal distribution when $n \to \infty$.

Proof. Condition 1) guarantees the uniform integrability of the
sequence μ_n, $n = 1, 2, \ldots$. Let us show that any converged
subsequence of μ_n, $n = 1, 2, \ldots$, is asymptotically normal and
by this proving Lemma 4.3.1. Suppose that $\mu_{n_k} \to \mu$ as $k \to \infty$
is some converged subsequence. Then we have

$$\lim_{k \to \infty} \int_{-\infty}^{\infty} (i\lambda - x) e^{i\lambda x} \mu_{n_k}(dx) = \int_{-\infty}^{\infty} (i\lambda - x) e^{i\lambda x} \mu(dx). \qquad (4.3.1)$$

Now if $\displaystyle f(\lambda) = \int_{-\infty}^{\infty} e^{i\lambda x} \mu(dx)$ then from (4.3.1) we obtain

$$i\lambda f(\lambda) = f'(\lambda) \frac{1}{i}$$

and hence

$$f'(\lambda) = -\lambda f(\lambda), \quad f(0) = 1.$$

Finally

$$f(\lambda) = e^{-\frac{\lambda^2}{2}}.$$

__Lemma 4.3.2.__ Let ξ_t, $t \in \mathbb{Z}^\nu$, be a s.r.f. such that with probability 1 $|\xi_t| < C$ and $DS_{I_n} \sim \sigma_0^2 \, |I_n|$, as $n \to \infty$, $\sigma_0^2 > 0$ for some increasing sequence of subsets

$$\ldots \subset I_n \subset I_{n+1} \subset \ldots \; , \qquad \bigcup_{n \in \mathbb{N}} I_n = \mathbb{Z}^\nu .$$

If for some sequence of numbers $p(n)$, $n \in \mathbb{N}$, such that

$$p^\nu(n) = 0\left(|I_n|^{1/2}\right), \quad n \to \infty$$

the following conditions are satisfied:

1)
$$\lim_{n \to \infty} \frac{1}{\sqrt{|I_n|}} \sum_{t \in I_n} E \xi_t \, e^{i\lambda\left(\frac{S_{I_n} - S_{n,t}}{\sigma_0 \sqrt{|I_n|}}\right)} = 0$$

2)
$$\lim_{n \to \infty} \left(\frac{1}{|I_n|} \sum_{t \in I_n} E \xi_t \frac{S_{n,t}}{\sigma_0 \sqrt{|I_n|}} \, e^{i\lambda\left(\frac{S_{I_n} - S_{n,t}}{\sigma_0 \sqrt{|I_n|}}\right)} - E \, e^{i\lambda \frac{S_{I_n}}{\sigma_0 \sqrt{|I_n|}}} \right) = 0$$

$$S_{n,t} = \sum_{s \in I_n : \, d(s,t) \leqslant p(n)} \xi_s \; , \qquad d(s,t) = \max_{1 \leqslant i \leqslant \nu} |t^{(i)} - s^{(i)}| .$$

Then for random field ξ_t , $t \in \mathbb{Z}^\nu$, the c.l.th. holds.

__Proof.__ Make sure that conditions of Lemma 4.3.1 hold. Since

$$DS_{I_n} \sim \sigma_0^2 \, |I_n| \; , \text{ as } \; n \to \infty \quad \text{then}$$
$$D\left(\frac{S_{I_n}}{\sigma_0 \sqrt{|I_n|}}\right) \leqslant C \, , \qquad C < \infty$$

and thus Condition 1) of Lemma 4.3.1 is valid.

Denote

$$\bar{S}_n = S_{I_n}\left(\sigma_0 \sqrt{|I_n|}\right)^{-1} , \; \bar{S}_{t,n} = S_{t,n}\left(\sigma_0 \sqrt{|I_n|}\right)^{-1} , \; \bar{\xi}_t = \xi_t \left(\sigma_0 \sqrt{|I_n|}\right)^{-1} .$$

To complete the proof it is sufficient to show that

$$\lim_{n \to \infty} E \, e^{i\lambda \bar{S}_n} \left(\bar{S}_n - i\lambda\right) = 0 .$$

We have

$$E \, e^{i\lambda \bar{S}_n} \left(\bar{S}_n - i\lambda\right) =$$

$$= \sum_{t \in I_n} E \bar{\xi}_t \, e^{i\lambda(\bar{S}_n - \bar{S}_{n,t})} \left(e^{i\lambda \bar{S}_{n,t}} - i\lambda \bar{S}_{n,t} - 1\right) + $$

$$\tag{4.3.2}$$

$$+ i\lambda \left(\sum_{t \in I_n} E \bar{\xi}_t \, \bar{S}_{n,t} \, e^{i\lambda(\bar{S}_n - \bar{S}_{n,t})} - E \, e^{i\lambda \bar{S}_n} \right) +$$

$$+ \sum_{t \in I_n} E \bar{\xi}_t \, e^{i\lambda(\bar{S}_n - \bar{S}_{n,t})} \, .$$

According to Conditions 1) and 2) of Lemma 4.3.2 the last terms on the right-hand side of (4.3.2) tends to zero as $n \to \infty$. It remains to verify that the first term in (4.3.2) tends to zero as $n \to \infty$. We have

$$\lim_{n \to \infty} \left| \sum_{t \in I_n} E \bar{\xi}_t \, e^{i\lambda(\bar{S}_n - \bar{S}_{t,n})} \left(e^{i\lambda \bar{S}_{t,n}} - i\lambda \bar{S}_{t,n} - 1 \right) \right| \leqslant$$

$$\leqslant \lim_{n \to \infty} C_1 |I_n|^{1/2} E (S_{t,n})^2 \leqslant C_2 \lim_{n \to \infty} \frac{p^\nu(n)}{|I_n|^{1/2}} = 0 \, .$$

Lemma 4.3.2 is proven.

4.4 Moments' (Semi-Invariants') Method

Let $\bar{\xi} = (\xi_1, \ldots, \xi_k)$ be a random vector,

$$\varphi_{\bar{\xi}}(t) = E \, e^{i(t, \bar{\xi})} = E \, e^{i \sum_{j=1}^{k} \xi_j t_j}, \quad t = (t_1, \ldots, t_k),$$

being its characteristic function. In case

$$E \, |\xi_i|^n < \infty, \quad i = \overline{1,k}$$

there exist moments $E \, \xi_1^{\nu_1} \ldots \xi_k^{\nu_k}$ for all non-negative integers $\nu_1, \ldots, \nu_k$ such that $\nu_1 + \cdots + \nu_k \leqslant n$. Hence both the existence and continuity of partial derivatives

$$\frac{\partial^{\nu_1 + \cdots + \nu_k}}{\partial t_1^{\nu_1} \ldots \partial t_k^{\nu_k}} \, \varphi_{\xi}(t_1, \ldots, t_k), \quad \nu_1 + \cdots + \nu_k \leqslant n$$

follow.

Expanding $\varphi_{\xi}(t_1, \ldots, t_k)$ in Taylor's series we have

$$\varphi_{\xi}(t_1, \ldots, t_k) = \sum_{\nu_1 + \cdots + \nu_k \leqslant n} \frac{i^{\nu_1 + \cdots + \nu_k}}{\nu_1! \ldots \nu_k!} \, m_{\xi}(\nu_1 \cdots \nu_k) t_1^{\nu_1} \ldots t_k^{\nu_k} + 0(|t|^n),$$

where $\quad |t| = |t_1| + |t_2| + \ldots + |t_k| \quad$ and

$$m_{\bar{\xi}}(\nu_1, \ldots, \nu_k) = E \xi_1^{\nu_1} \ldots \xi_k^{\nu_k}.$$

The function $\varphi_\xi(t_1, \ldots, t_k)$ is continuous, $\varphi_\xi(0, \ldots, 0) = 1$ and therefore in some neighbourhood of zero $\quad |t| < \delta \quad$ it does not vanish. In this neighbourhood there exist continuous partial derivatives

$$\frac{\partial^{\nu_1 + \ldots + \nu_k}}{\partial t_1^{\nu_1} \ldots \partial t_k^{\nu_k}} \, \ell n \, \varphi_\xi(t_1, \ldots, t_k).$$

That's why

$$\ell n \, \varphi_\xi(t_1, \ldots, t_k) = \sum_{\nu_1 + \ldots + \nu_k \leqslant n} \frac{i^{\nu_1 + \ldots + \nu_k}}{\nu_1! \ldots \nu_k!} \, S_\xi^{(\nu_1, \ldots, \nu_k)} t_1^{\nu_1} \ldots t_k^{\nu} + O(|t|^n),$$

where the coefficients $\quad S_\xi^{(\nu_1, \ldots, \nu_k)} \quad$ are called mixed semi-invariants of order $\quad \nu = (\nu_1, \ldots, \nu_k)$.

Let

$$\nu! = \nu_1! \ldots \nu_k! \, , \qquad |\nu| = \nu_1 + \cdots + \nu_k \, ,$$

$$S_\xi^{(\nu)} = S_\xi^{(\nu_1, \ldots, \nu_k)} \, , \qquad m_\xi^{(\nu)} = m_\xi^{(\nu_1, \ldots, \nu_k)}.$$

Hence the relation between semi-invariants and moments is expressed with the help of the following finite formulas:

$$m_\xi^{(\nu)} = \sum_{\lambda^{(1)} + \ldots + \lambda^{(q)} = \nu} \frac{1}{q!} \frac{\nu!}{\lambda^{(1)}! \ldots \lambda^{(q)}!} \prod_{p=1}^{q} S_\xi^{(\lambda^{(p)})} \, , \qquad (4.4.1)$$

$$S_\xi^{(\nu)} = \sum_{\lambda^{(1)} + \ldots + \lambda^{(q)} = \nu} \frac{(-1)^{q-1}}{q} \frac{\nu!}{\lambda^{(1)} \ldots \lambda^{(q)}} \prod_{p=1}^{q} m_\xi^{(\lambda^{(p)})} \, ,$$

where the summation is taken over all ordered sets of integer non-negative nonempty vectors $\lambda^{(p)}$, $|\lambda^{(p)}| > 0$ giving vector ν as a sum.

Let $\quad m_\xi^{(k)} = m^{(k)} = E \xi^k \quad$ and $\quad S^{(k)} = S_\xi^{(k)} \quad$ respectively. The following theorem lies in the basis of the application of semi-invariants technique for limit theorems.

<u>Theorem 4.4.1</u> (Frechet, Shochat). Let ξ_n, $n \in \mathbb{N}$, be a sequence of random variables and

$$m_n^{(\kappa)} = m_{\xi_n}^{(\kappa)} < \infty, \qquad F_n(x) = P(\xi_n < x), \quad n \in \mathbb{N}.$$

Suppose that there exist the finite limits

$$\lim_{n \to \infty} m_n^{(\kappa)} = m^{(\kappa)} \qquad\qquad \text{for all } k \in \mathbb{N}.$$

Then the limits $m^{(k)}$ represent the moments of distribution function $F(x)$ such that there exists a subsequence of $F_n(x)$ $n \in \mathbb{N}$ weakly converged to $F(x)$. If in addition these limits uniquely determine $F(x)$ then $F_n(x)$ converges weakly to

$$F(x).$$

In proving the c.l.th. by Moments' method usually it is more **convenient** to verify the convergence of semi-invariants. Let ξ be a normally distributed with mean 0 and variance 1. Then

$$S_\xi^{(2)} = 1, \qquad S_\xi^{(\kappa)} = 0, \qquad k \in \mathbb{N}, \; k \neq 2$$

and according to Theorem 4.4.1 and relations (4.4.1) the sequence $F_n(x)$ converges to the standard normal law if

$$S_{\xi_n}^{(\kappa)} \to 0, \qquad n \to \infty, \qquad k \neq 0$$

and

$$S_{\xi_n}^{(2)} \to 1, \qquad n \to \infty.$$

This method is widely used in statistical physics since there is **developed technique (cluster expansions) which gives the possi**bility to obtain the appropriate estimates for semi-invariants of Gibbs random fields (see Chapter 9).

Note, that the applications of semi-invariants techniques to dependent random variables are presented in /80/, /118/, /136/, /138/.

In this chapter for a s.r.p. ξ_t, $t \in \mathbb{Z}^1$, satisfying the above mentioned mixing conditions we will study a limit behaviour of the sums

$$\frac{1}{B_n} \sum_{t=1}^{n} \xi_t - A_n . \qquad (5.0.1)$$

Here A_n, $n \in \mathbb{N}$ and B_n, $n \in \mathbb{N}$ are some sequences of real numbers.

Let us formulate one general theorem which shows that the limit distributions for the sequences of (5.0.1) belong to the class of stable law.

Theorem 5.0.1(/66/). Let ξ_t, $t \in \mathbb{Z}^1$, be a α -mixing s.r.p. If $B_n \to \infty$, as $n \to \infty$ and the distribution functions $F_n(x)$ of sums (5.0.1) converge weakly to the distribution function $F(x)$, then the latter is a stable one. Moreover, if the exponent of the limit distribution is equal to γ , then

$$B_n = n^{\frac{1}{\gamma}} h(n) , \quad n \in \mathbb{N}, \quad 0 < \gamma \leq 2 ,$$

where $h(n)$ is a slowly varying function.

5.1 Central Limit Theorem for α -mixing Stationary

Random Processes

In the beginning we will give some necessary and sufficient conditions for the validity of the c.l.th.

Theorem 5.1.1 (/66/). Let ξ_t, $t \in \mathbb{Z}^1$, be an α -mixing s.r.p., $E\xi_0 = 0$, $E\xi_0^2 < \infty$. For the c.l.th. to be valid for this process and $\sigma_n^2 \to \infty$, as $n \to \infty$ it is necessary that:

1) $\sigma_n^2 = n h(n)$ where $h(n)$ is a slowly varying function defined on R_+^1 ;

2) for any pair of natural valued functions $p = p(n)$ and $q = q(n)$ such that:

   a) $p \to \infty$, $q \to \infty$, $p = o(n)$, $q = o(p)$ as $n \to \infty$;

   b) $\lim\limits_{n \to \infty} p^{-2} n^{1-\beta} q^{1+\beta} = 0$ for some $\beta > 0$;   (5.1.1)

   c) $\lim\limits_{n \to \infty} p^{-1} n \alpha(q) = 0$;

3) for any $\varepsilon > 0$

$$\lim_{n \to \infty} \int_{|S_p| \geqslant \varepsilon \sigma_n} \left(\frac{S_p}{\sigma_p}\right)^2 P(d\omega) = 0. \qquad (5.1.2)$$

Conversely, if the condition 1) is valid and (5.1.2) is fulfilled for some pair of functions p, q as in (5.1.1), then the s.r.p. ξ_t, $t \in \mathbb{Z}^1$, satisfies the c.l.th.

<u>Proof.</u> Let us show the necessity of Condition 1). According to Remark 3.1.4 it is sufficient to prove that the sequence $\left(\frac{S_n}{\sigma_n}\right)^2$ is uniformly integrable. Since the distribution function
$$F_n(x) = P\left(\frac{S_n}{\sigma_n} < x\right)$$
converges weakly to standard normal distribution function $\Phi(x) = P(\xi < x)$ then for any fixed $C > 0$ we have

$$\lim_{n \to \infty} \int \left(\frac{S_n}{\sigma_n}\right)^2 I\left(\left|\frac{S_n}{\sigma_n}\right|^2 < C\right) P(d\omega) = \int \xi^2 I(\xi^2 < C) P(d\omega).$$

Further

$$\lim_{n \to \infty} \int \left(\frac{S_n}{\sigma_n}\right)^2 I\left(\left|\frac{S_n}{\sigma_n}\right|^2 > C\right) P(d\omega) = 1 -$$

$$- \lim_{n \to \infty} \int \left(\frac{S_n}{\sigma_n}\right)^2 I\left(\left|\frac{S_n}{\sigma_n}\right|^2 < C\right) P(d\omega) = 1 - \int \xi^2 I(\xi^2 \leqslant C) P(d\omega) =$$

$$= \int \xi^2 I(\xi^2 > C) P(d\omega).$$

Evidently it implies the uniform integrability of the sequence
$$\left(\frac{S_n}{\sigma_n}\right)^2, \quad n \in \mathbb{N}.$$

Let p, q be some pair of functions, satisfying Condition 2) of Theorem 5.1.1. For example, one can take the following pair:

$$p = max\left\{\left[\frac{n\,\alpha([n^{1/4}])}{\lambda(n)}\right], \left[\frac{n^{3/4}}{\lambda(n)}\right]\right\}, \quad q = [n^{1/4}],$$

$$\lambda(n) = max\left\{\alpha([n^{1/4}])^{1/3}, \frac{1}{\ln n}\right\}.$$

Here $[\cdot]$ is the integer function.

Let us represent the sum S_n in a standard form:

$$S_n = \sum_{i=0}^{K-1} X_i + \sum_{i=0}^{K} Y_i = S_n' + S_n'', \qquad (5.1.3)$$

$$X_i = \sum_{j=i(p+q)+1}^{i(p+q)+p} \xi_j \ , \qquad i = 0, 1, \ldots, k-1 \ ,$$

$$Y_i = \sum_{j=i(p+q)+p+1}^{(i+1)(p+q)} \xi_j \ , \qquad i = 0, 1, \ldots, k-1 \ ,$$

$$Y_k = \sum_{j=k(p+q)+1}^{n} \xi_j \ , \qquad k = k(n) \approx \left[\frac{n}{p+q} \right]$$

and show that by the given choice of the pair p, q the limiting behaviour of the sequence $\sigma_n^{-1} S_n$, $n \in \mathbb{N}$, is determined by that of the sequence $\sigma_n^{-1} S_n'$, $n \in \mathbb{N}$. For this it is sufficient to prove that

$$P\left(\left| \frac{S_n''}{\sigma_n} \right| \geqslant x \right) \longrightarrow 0, \qquad n \to \infty.$$

According to Chebishev inequality we have

$$P\left(\left| \frac{S_n''}{\sigma_n} \right| \geqslant x \right) \leqslant (x \sigma_n)^{-2} DS_n'' \ , \qquad x > 0$$

and

$$\frac{DS_n''}{\sigma_n^2} \leqslant \frac{k^2 q h(q)}{n h(n)} + \frac{2k}{n h(n)} \sqrt{q h(q) q' h(q')} + \frac{q' h(q')}{n h(n)} \ . \qquad (5.1.4)$$

Let us show that all summands on the right-hand side of (5.1.4) tend to zero as $n \to \infty$. Indeed,

$$\frac{k^2 q h(q)}{n h(n)} \sim \frac{n q h(q)}{p^2 h(n)} = \left[\left(\frac{q}{n} \right)^\beta \frac{h\left(n \frac{q}{n} \right)}{h(n)} \right] \frac{n^{1-\beta} q^{1+\beta}}{p^2} \leqslant$$

$$\leqslant C \left(\frac{q}{n} \right)^\beta \left(\frac{n}{q} \right)^\beta \cdot \frac{n^{1-\beta} q^{1+\beta}}{p^2} \longrightarrow 0, \qquad n \to \infty.$$

Here we use Property 4) of a slowly varying function (see Chapter 1). Taking into consideration that $q' \leqslant p + q$ and using Property 4) once again we get that the other terms on the right-hand side of (5.1.4) tend to zero, as $n \to \infty$.
Further using the inequality (3.3.5) and Lemma 2.1.1.
we obtain

$$\left| E \prod_{j=1}^{k} e^{it \frac{X_j}{\sigma_n}} - \prod_{j=1}^{k} E e^{it \frac{X_j}{\sigma_n}} \right| \leqslant k \alpha(q) \ . \qquad (5.1.5)$$

It follows from the properties of the functions p, q: $k \alpha(q) \to 0$ as $n \to \infty$. This implies that the limiting distribution of

$\sigma_n^{-1} S_n^1$ coincides with that of the sums $\sum_{j=1}^{K} Z_{jn}$, where Z_{jn} $j=1,\ldots,K(n)$ is the double sequence of independent random variables each of which in n-th series is identically distributed with the random variable $\sigma_n^{-1} X_1$. For $\sum_{j=1}^{K} Z_{jn}$ to be asymptotically normal it is necessary and sufficient that the Lindeberg condition is fulfilled (i.e. Condition 3) of Theorem 5.1.1).

<u>Theorem 5.1.2 (/28/,/88/).</u> Let ξ_t, $t \in \mathbb{Z}^1$ be a centered α-mixing s.r.p. with finite second moments and

$$\sigma_n^2 \to \infty, \quad n \to \infty$$

Then the c.l.th. is valid for this process iff the sequence $\sigma_n^{-2} S_n^2$, $n \in \mathbb{N}$ is uniformly integrable.

<u>Proof.</u> Let us prove this theorem under condition

$$\sigma_n^2 = n h(n),$$

where $h(n)$ is a slowly varying function defined on R_+^1. As Theorem 3.1.5 shows this assumption does not break the generality. The necessity of the conditions of this theorem was practically obtained in the proof of Theorem 5.1.1. Let us prove their sufficiency. We will show that any convergent subsequence $\sigma_{f(n)}^{-1} S_{f(n)}$, $f(n) \uparrow \infty$, $n \to \infty$ contains an asymptotically normal subsequence. For this it is suffi‐ cient to prove that for some standard pair of functions p, q satis‐ fying (5.1.1) the following relation is valid

$$\lim_{n \to \infty} \int (\sigma_{pf}^{-1} S_{pf})^2 I \left(|\sigma_{pf}^{-1} S_{pf}| > \varepsilon \sigma_{pf}^{-1} \sigma_f \right) P(d\omega) = 0,$$

$$pf = p(f(n)), \quad n \in \mathbb{N}, \quad \varepsilon > 0.$$

The sequence $\sigma_{pf}^{-1} S_{pf}$, $n \in \mathbb{N}$ itself is a subsequence (possibly with repeating terms) of the sequence $\sigma_n^{-1} S_n$, $n \in \mathbb{N}$. Without los‐ ing of generality we can assume that it is convergent and let F be its limiting distribution function. Since $\sigma_{pf}^{-1} S_{pf}$, $n \in \mathbb{N}$ is uni‐ formly integrable we can write

$$\lim_{n \to \infty} \int (\sigma_{pf}^{-1} S_{pf})^2 I \left(|\sigma_{pf}^{-2} S_{pf}^2| > \varepsilon \sigma_f^2 \sigma_{pf}^{-2} \right) P(d\omega) \leq$$

$$\leq \lim_{n \to \infty} \int (\sigma_{pf}^{-1} S_{pf})^2 I \left(|\sigma_{pf}^{-2} S_{pf}^2| > C \right) P(d\omega) = \int_{x^2 > C} x^2 dF(x).$$

As far as the constant C is arbitrary the theorem is proved. The next theorem is the evident corollary of the previous one.

__Theorem 5.1.3.__ Let ξ_t, $t \in \mathbb{Z}^1$ be a centered α -mixing s.r.p., $E\xi_0^2 < \infty$ and

$$\sigma_n^2 \to \infty, \quad n \to \infty .$$

Suppose that for some standard pair of functions p, q satisfying the condition (5.1.1) the sequence $\sigma_p^{-1} S_p$, $n \in \mathbb{N}$ is asymptotically normal, then the sequence $\sigma_n^{-1} S_n$, $n \in \mathbb{N}$ itself is asymptotically normal.

In connection with Herrndorf examples (see Chapter 2 of this book) we will prove the following theorem.

__Theorem 5.1.4.__ Suppose that ξ_t, $t \in \mathbb{Z}^1$ is an α -mixing s.r.p.; $E\xi_0 = 0$, $E\xi_0^2 < \infty$ and $\sigma_n^2 = n h(n)$, where $h(n)$ is a slowly varying function on R_+^1. Then there exists an asymptotically Gaussian subsequence of the sequence $\sigma_n^{-1} S_n$ with variance $\beta, 0 \leqslant \beta \leqslant 1$ (the case $\beta = 0$ corresponds to the degenerate distribution).

In the beginning we prove another theorem.

__Theorem 5.1.5.__ Suppose that a s.r.p. ξ_t, $t \in \mathbb{Z}^1$ is centered and α -mixing. Let $E\xi_0^2 < \infty$ and

$$\sigma_n^2 = n h(n) ,$$

where $h(n)$ is a slowly varying function defined on R_+^1. Then the limiting distribution function of any convergent subsequence $\sigma_f^{-1} S_f$, $n \in \mathbb{N}$ of the sequence $\sigma_n^{-1} S_n$, $n \in \mathbb{N}$ is a stable one.

__Proof.__ Here we use the analogy whiththe proof of Theorem 5.0.1. Let $\sigma_f^{-1} S_f$ be some convergent subsequence of $\sigma_n^{-1} S_n$ and let $F(x)$ be its limiting distribution function. Suppose that it is non-degenerate (the degenerate distribution function is stable). Since $\sigma_n^2 = n h(n)$, then for any positive constants a_1 and a_2 there exists a sequence $m(n)$, $n \in \mathbb{N}$ such that

$$\lim_{n \to \infty} \sigma_n^{-1} \sigma_{m(n)} = a_2^{-1} a_1 .$$

Obviously

$$\lim_{n \to \infty} \sigma_f^{-1} \sigma_{mf} = a_2^{-1} a_1 , \qquad mf = m(f(n)) .$$

Let $\tau(n)$ $(\tau(n) \to \infty, \; n \to \infty)$ be a function such that in probability

$$\sigma_n^{-1} \sum_{j=1}^{\tau} \xi_j \to 0, \quad n \to \infty.$$

Consider the sum

$$\frac{1}{a_1}\left(\frac{1}{\sigma_f}\sum_{j=1}^{f}\xi_j - 1\right) + \frac{\sigma_{mf}}{a_1\sigma_f}\left(\frac{1}{\sigma_{mf}}\sum_{j=f+\tau f+1}^{f+\tau f+mf}\xi_j - \beta_2\right) =$$

$$= \left(\frac{1}{a_1\sigma_f}\sum_{j=1}^{f+\tau f+mf}\xi_j - C_f\right) - \frac{1}{a_1\sigma_f}\sum_{j=f+1}^{f+\tau f}\xi_j, \tag{5.1.6}$$

$$C_f = -\frac{\beta_1}{a_1} - \frac{\beta_2}{a_1}\frac{\sigma_{mf}}{\sigma_f}.$$

Using Lemma 2.1.1 it is easy to see that the limit of the left-hand side of (5.1.6) is equal to the following convolution

$$F(a_1 x + \beta_1) * F(a_2 x + \beta_2).$$

In its turn the limit of the right-hand side of (5.1.6) is equal to $F(ax + \beta)$,for some $a > 0$ and $\beta \in R^1$. Theorem 5.1.5 is proved.

<u>Proof of Theorem 5.1.4.</u> Here we will use the following well-known result of Khinchin/134/: Let

$$X_{n1}, \; X_{n2}, \; \ldots, \; X_{n K_n}$$

$$K_n \to \infty, \quad n \to \infty$$

be a triangular array of random variables independent in each series. If the sum $\sum_{j=1}^{K_n} X_{nj}$ is weakly convergent and

$$\sum_{j=1}^{K_n} P(X_{nj} \geq \varepsilon) \to 0, \quad n \to \infty, \; \varepsilon > 0$$

then the limiting distribution is Gaussian with variance β, $0 \leq \beta < \infty$ (the case $\beta = 0$ corresponds to the degenerate distribution function) Let p, q be some standard pair of functions satisfying (5.1.5). The sequence $\sigma_p^{-1} S_p$, $n \in \mathbb{N}$ is a subsequence of $\sigma_n^{-1} S_n$ and let $\sigma_{pf}^{-1} S_{pf}$ be some of its convergent subsequences. According to Theorem 5.1.4 the limiting distribution function $F(x)$ for this subsequence is stable and hence it is continuous. Obviously, there exists some function $qf = q(f(n)) \to \infty$ as $n \to \infty$ such that

$$pf = 0(qf), \quad n \to \infty, \qquad \frac{qf}{pf} \alpha(qf) \to 0, \quad n \to \infty$$

$$\frac{qf}{pf} \sup_{x} |P(\sigma_{pf}^{-1} S_{pf} < x) - F(x)| \to 0, \quad n \to \infty. \tag{5.1.6'}$$

Let us now consider the sequence $\sigma_{qf}^{-1} S_{qf}$, $n \in \mathbb{N}$. Without losing generality we can assume that it is convergent. Using Bernstein's procedure to this subsequence and taking into account (5.1.6') we obtain that for corresponding triangulare array the conditions of Khinchin's theorem are valid. The theorem is proved.

The direct consequence of this theorem is the following result recently obtained in /26/.

<u>Theorem 5.1.6.</u> Suppose that ξ_t, $t \in \mathbb{Z}^1$ is a centered α -mixing s.r.p. with finite second moments and

$$\sigma_n^2 = n h(n),$$

where $h(n)$ is a slowly varying function on R_+^1. Then the condition

$$\varliminf_{n \to \infty} \frac{E^{\frac{1}{2}}|S_n|^2}{E|S_n|} \le \sqrt{\frac{\pi}{2}} \tag{5.1.6''}$$

is necessary and sufficient for the validity of c.l.th.

<u>Proof.</u> The necessity is evident since for asymptotically normal sequences $\sigma_n^{-1} S_n$, $n \in \mathbb{N}$ we have

$$\lim_{n \to \infty} E\left|\frac{S_n}{\sigma_n}\right| = \sqrt{\frac{\pi}{2}}.$$

Concerning the sufficiency of the conditions let us show that any
convergent subsequence $\sigma_f^{-1} S_f$, $n \in \mathbb{N}$ contains an asymptotically normal subsequence. In exactly the _same_ way as in the proof
of Theorem 5.1.5 we can construct the asymptotically Gaussian subsequence $\sigma_{gf}^{-1} S_{gf}$, $n \in \mathbb{N}$ with variance β, $0 \leqslant \beta \leqslant 1$. By
the condition (5.1.6") it is asymptotically normal since

$$\varlimsup_{n \to \infty} \frac{E^{\frac{1}{2}} |S_n|^2}{E |S_n|} = \frac{1}{\beta \sqrt{\frac{2}{\pi}}} \leqslant \sqrt{\frac{\pi}{2}}$$

and hence $\beta = 1$.
It remains to use Theorem 5.1.3.

Note, that the conditions of the above mentioned theorems are
difficult to check. Now we give the classical Ibragimov's theorem
in which the sufficient conditions for c.l.th. are formulated in
terms of the moments and the rate of mixing.

<u>Theorem 5.1.7 (/66/)</u>. Let ξ_t, $t \in \mathbb{Z}^1$ be a centered, α-mixing s.r.p. satisfying one of the following two conditions:

1) for some $\delta > 0$, $E |\xi_0|^{2+\delta} < \infty$ and $\sum_{n=1}^{\infty} \alpha^{\frac{\delta/2}{2+\delta}}(n) < \infty$;
or

2) with probability 1, $|\xi_0| < C$ and $\sum_{n=1}^{\infty} \alpha(n) < \infty$.
Then $\sigma^2 = E \xi_0^2 + 2 \sum_{K=1}^{\infty} E \xi_0 \xi_K < \infty$ and if $\sigma^2 \neq 0$ the
c.l.th. for the process ξ_t, $t \in \mathbb{Z}^1$ holds.

Now we will give different proofs of this theorem to demonstrate
the possibilities of the methods described in the previous chapter
and also in a view that some of the used mathematical steps will be
necessary later on.

<u>1) Proof by using "small" blocks.</u>

According to Lemma 4.1.3 it is sufficient to check that for any
standard pair of functions p, q the following relation
holds:

89

$$\left| E \prod_{j=1}^{k} e^{itS(n,j)} - \prod_{j=1}^{k} E\, e^{itS(n,j)} \right| \to 0, \quad n \to \infty, \qquad (5.1.7)$$

$$S(n,j) = \sigma_n^{-1}\, S_{\Delta_j^{(n)}}.$$

Note that the condition $\sigma_n^2 \sim \sigma^2 n$, $n \to \infty$, $\sigma^2 > 0$ is valid by virtue of Theorem 3.1.4. Using the inequalities (3.3.7) and Lemma 2.1.1 we get:

$$\mathfrak{m}_k^t \leq \sum_{j=1}^{k-1} \sum_{m=j+1}^{k} \left| E\left(e^{itS(n,j)} - 1\right)\left(e^{itS(n,m)} - 1\right) \prod_{s=m+1}^{k} e^{itS(n,s)} - \right.$$

$$\left. - E\left(e^{itS(n,j)} - 1\right) E\left(e^{itS(n,m)} - 1\right) \prod_{s=m+1}^{k} e^{itS(n,s)} \right| \leq$$

$$\leq C\, |t|^2\, \frac{n}{\sigma_n^2 p}\, E^{\frac{2}{2+\delta}}\, \left| \sum_{j=1}^{P} \xi_j \right|^{2+\delta} \sum_{j=1}^{k} \alpha^{\frac{\delta/2+\delta}{}}(iq).$$

Applying the Minkowski inequality we obtain

$$\mathfrak{m}_k^t \leq C_1\, |t|^2\, p\, E^{\frac{2}{2+\delta}} |\xi_0|^{2+\delta} \sum_{j=1}^{k} \alpha^{\frac{\delta/2+\delta}{}}(jq).$$

The monotonicity of the mixing coefficient implies

$$\alpha^{\frac{\delta/2+\delta}{}}(jq) \leq \frac{2}{q} \sum_{k \geq (j-\frac{1}{2})q}^{jq} \alpha^{\frac{\delta/2+\delta}{}}(k).$$

Hence

$$\sum_{j=1}^{k} \alpha^{\frac{\delta/2+\delta}{}}(jq) \leq \frac{2}{q} \sum_{k \geq q/2}^{\infty} \alpha^{\frac{\delta/2+\delta}{}}(k). \qquad (5.1.8)$$

Finally

$$\mathfrak{m}_k^t \leq C_2\, |t|^2\, \frac{p}{q} \sum_{j \geq q/2}^{\infty} \alpha^{\frac{\delta/2+\delta}{}}(j)$$

and it is clear that for any function $p = p(n)$, $p(n) \to \infty$, $p = o(n)$ as $n \to \infty$ one can choose a function $q = q(n)$, $q(n) \to \infty$, $q = o(p)$ as $n \to \infty$ and such that

$$\mathfrak{m}_k^t \to 0 \qquad \text{when} \quad n \to \infty.$$

For the case of bounded random variables Theorem 5.1.7 is proved analogously.

2) Proof by Gordin's method (/60/)

Let us check the validity of the conditions of Theorem 4.2.3.
First consider the case of bounded random variables. Using
Lemma 2.1.1 we can write

$$\sum_{k=1}^{\infty} |E(\xi_k E(\xi_N | \mathfrak{N}_{-\infty}^0))| = \sum_{k=1}^{N} |E(\xi_k E(\xi_N | \mathfrak{N}_{-\infty}^0))| +$$

$$+ \sum_{k=1}^{\infty} |E(\xi_{k+N} E(\xi_N / \mathfrak{N}_{-\infty}^0))| \leq$$

$$\leq C_1 \sum_{k=1}^{N} \left(E(E(\xi_0 / \mathfrak{N}_{-\infty}^{-N}))^2\right)^{1/2} \alpha^{1/2}(k) +$$

$$+ C_2 \sum_{k=1}^{N} \alpha(k+N) \leq C_3 \alpha^{\frac{1}{2}}(N) \sum_{k=1}^{N} \alpha^{\frac{1}{2}}(k) + C_2 \sum_{k=1}^{\infty} \alpha(k+N).$$

By the monotonicity of the mixing coefficient and Condition 2)
of Theorem 5.1.7 we get

$$k\alpha(k) \to 0, \quad k \to \infty, \qquad\qquad k\alpha(k) < A < \infty,$$

for all $k \in \mathbb{N}$.
Thus

$$\alpha^{1/2}(N) \sum_{k=1}^{N} \alpha^{1/2}(k) < A^{1/2} \alpha^{1/2}(N) \sum_{k=1}^{N} k^{-1/2} <$$

$$< C_4 (N\alpha(N))^{1/2} \to 0, \quad N \to \infty.$$

In the case of unbounded summands by using the estimates of
Lemma 2.1.1 again we can write

$$\sum_{k=1}^{\infty} |E(\xi_k E(\xi_N | \mathfrak{N}_{-\infty}^0))| \leq C_1 (E|\xi_0|^{2+\delta})^{\frac{1}{2+\delta}} \times$$

$$(E|E(\xi_0 / \mathfrak{N}_{-\infty}^{-N})|^{2+\delta})^{\frac{1}{2+\delta}} \sum_{k=1}^{\infty} \alpha^{\frac{\delta}{2+\delta}}(k).$$

Let us show that

$$E|E(\xi_0 / \mathfrak{N}_{-\infty}^{-N})|^{2+\delta} \to 0, \quad \text{as} \quad N \to \infty.$$

By Jensen's inequality we have

$$\left| E\left(\xi_0 / \mathcal{O}_{-\infty}^{-N}\right)\right|^{2+\delta} \le E\left(|\xi_0|^{2+\delta} / \mathcal{O}_{-\infty}^{-N}\right).$$

Hence

$$E\left(E\left(|\xi_0|^{2+\delta} / \mathcal{O}_{-\infty}^{-N}\right)\right) = E|\xi_0|^{2+\delta} < \infty$$

and from the martingale theorem of convergence it follows

$$E\left(\xi_0 / \mathcal{O}_{-\infty}^{-N}\right) \to 0 \qquad \text{a.s. when} \quad n \to \infty$$

(since the α -mixing condition implies $\bigcap_{N=1}^{\infty} \mathcal{a}_{-\infty}^{-N} = \mathcal{U}$, where $\mathcal{U}$ is trivial σ -algebra).

<u>3) Classical proof.</u> We will give the proof only in case of the unbounded random variables, assuming that the c.l.th. holds for the bounded ones. The truncating method considered here will be used later on. Denote

$$f_N(x) = \left\{ \begin{array}{ll} x, & |x| \le N, \\ 0, & |x| > N. \end{array} \right.$$

By Proposition 2.3.1   the sequence of random variables $\eta_t^{(N)} = f_N(\xi_t),\ t \in \mathbb{Z}^1,$ for fixed N is a stationary and α -mixing at least with the same mixing coefficient. Obviously

$$\left|\eta_t^{(N)}\right| \le N, \quad t \in \mathbb{Z}^1.$$

Besides, we have

$$E\eta_t^{(N)} \to 0, \quad E\eta_0^{(N)}\eta_t^{(N)} \to E\xi_0\xi_t , \text{ as } \qquad N \to \infty.$$

Hence it follows for sufficiently large N

$$\sigma^2(N) = E\left(\eta_0^{(N)} - E\eta_0^{(N)}\right)^2 + 2\sum_{j=1}^{\infty} E\left(\eta_0^{(N)} - E\eta_0^{(N)}\right)\left(\eta_j^{(N)} - E\eta_j^{(N)}\right) > \frac{\sigma^2}{2} > 0.$$

For such N we have

$$\sigma_n^2(N) = E\left(\sum_{j=1}^{n}\left(\eta_j^{(N)} - E\eta_j^{(N)}\right)\right)^2 = n\sigma^2(N)(1 + 0(1)) >$$

$$> \frac{\sigma^2 n}{2}(1 + 0(1)) \to \infty , \quad \text{as} \qquad n \to \infty .$$

Now we will prove the uniform integrability of the sequence

$$\sigma_n^{-2}(N)\left(\sum_{j=1}^{n}\left(\eta_j^{(N)} - E\eta_j^{(N)}\right)\right)^2, \quad n \in \mathbb{N},$$

for fixed N . This implies that the sequence

$$\sigma_n^{-1}(N)S_n^{(N)} = \sigma_n^{-1}(N)\sum_{j=1}^{n}(\eta_j^{(N)} - E\eta_j^{(N)})^2 \quad \text{is asymptotically normal as}$$

$n \to \infty$ for each N .

Using Proposition 3.2.1 we have

$$E\left(\frac{S_n^{(N)}}{\sigma_n(N)}\right)^2 I\left(\left|\frac{S_n^{(N)}}{\sigma_n(N)}\right| \geqslant K\right) \leqslant K^{-\delta} E\left|\frac{S_n^{(N)}}{\sigma_n(N)}\right|^{2+\delta} \leqslant CK^{-\delta},$$
$$0 < C < \infty, \quad K > 0 .$$

Thus for any fixed and sufficiently large N the sequence $\sigma_n^{-1}(N)\, S_n^{(N)}$ is asymptotically normal.

Next we can write

$$z_n = \frac{S_n}{\sigma\sqrt{n}} = \frac{\sum_{j=1}^{n}(\eta_j^{(N)} - E\eta_j^{(N)})}{\sigma(N)\sqrt{n}} \cdot \frac{\sigma(N)}{\sigma} + \frac{\sum_{j=1}^{n}(\bar{\eta}_j^{(N)} - E\bar{\eta}_j^{(N)})}{\sigma\sqrt{n}} =$$
$$= z_n' + z_n'' .$$

Here $\bar{\eta}_j^{(N)} = \eta_j - \eta_j^{(N)} .$

Let us show that z_n'' in probability tends to zero as $n \to \infty$. Indeed,

$$(\sigma\sqrt{n})^{-2} E\left(\sum_{j=1}^{n}(\bar{\eta}_j^{(N)} - E\bar{\eta}_j^{(N)})\right)^2 = \sigma^{-2}\left(E(\bar{\eta}_j^{(N)} - E\bar{\eta}_j^{(N)})^2 + \right.$$

$$\left. + \frac{2}{n}\sum_{j=1}^{n}(n-j)E(\bar{\eta}_0^{(N)} - E\bar{\eta}_0^{(N)})(\bar{\eta}_j^{(N)} - E\bar{\eta}_j^{(N)})\right) \leqslant$$

$$\leqslant C_1 E|\bar{\eta}_0^{(N)}|^2 + C_2 \sum_{j=N+1}^{\infty}\alpha^{\frac{\delta}{2+\delta}}(j) \to 0, \quad \text{as} \quad N \to \infty.$$

For the given $\varepsilon > 0$ let's choose N such that

$$z_n'' \leqslant \varepsilon \quad \text{and} \quad |1 - \sigma^{-1}\sigma(N)| \leqslant \varepsilon .$$

Hence

$$\left|e^{z_n} - e^{-\frac{t^2}{2}}\right| \leqslant \left|E e^{it z_n'} - e^{-\frac{t^2}{2}}\right| +$$

$$+ |E\, e^{it z_n'}(e^{it z_n''} - 1)| \le |e^{-\frac{\sigma^2(N)}{\sigma^2}\frac{t^2}{2}} - e^{-\frac{t^2}{2}}| +$$

$$+ E|e^{it z_n''} - 1| + |E\, e^{it z_n'} - e^{-\frac{\sigma^2(N)}{\sigma^2}\frac{t^2}{2}}| \le$$

$$\le \frac{t^2}{2}|1 - \frac{\sigma^2(N)}{\sigma^2}| + t\, E|z_n''| + o(1) \le$$

$$\le \varepsilon t^2 + t\sqrt{\varepsilon} + o(1), \qquad n \to \infty.$$

The theorem is proven.

Looking through this proof one can see that the reduction to the bounded random variables requires mixing conditions only of the following type:

$$\alpha_{s,t} = \alpha(\sigma(\xi_t), \sigma(\xi_s)), \qquad s,t \in \mathbb{Z}^1.$$

Therefore a stronger result is valid.

__Theorem 5.1.8.__ Let ξ_t, $t \in \mathbb{Z}^1$, be a centered α -mixing s.r.p. and for some $\delta > 0$, $E|\xi_0|^{2+\delta} < \infty$. If

$$\sum_{n=1}^{\infty} \alpha(n) < \infty \qquad \text{and} \qquad \sum_{n=1}^{\infty} \alpha_{0,n}^{\delta/2+\delta} < \infty,$$

then $\sigma^2 = E\xi_0^2 + 2\sum_{j=1}^{\infty} E\xi_0\xi_j < \infty$ and if $\sigma^2 > 0$ the c.l.th. for the process ξ_t, $t \in \mathbb{Z}^1$, holds.

As it was shown in Chapter 3 the variance of the sums in Davydov's examples has no standard behaviour, i.e., $DS_n \neq n h(n)$, where $h(n)$ is a slowly varying function.

Now according to Theorem 5.0.1 such processes are not subjected to the c.l.th. Thus the following facts hold.

__Theorem 5.1.9.__ One can construct such s.r.p. ξ_t, $t \in \mathbb{Z}^1$, that:

1) For any $\varepsilon, \delta > 0$ $\quad E|\xi_0|^{2+\delta} < \infty$, $\quad \sum_{n=1}^{\infty} \alpha^{\frac{\delta}{2+\delta}+\varepsilon}(n) < \infty$

but
$$\sigma_n^{-1} S_n \not\to N(0,1), \qquad n \to \infty$$

or

2) For any $\delta > \varepsilon > 0$ $\quad E|\xi_0|^{2+\delta-\varepsilon} < \infty$, $\quad \sum_{n=1}^{\infty} \alpha^{\frac{\delta/2+\delta}{}}(n) < \infty$

but again

$$\sigma_n^{-1} S_n \not\to N(0,1), \qquad\qquad n \to \infty.$$

3) For any $\quad \varepsilon > 0 \quad |\xi_0| < C \qquad$ with probability 1

$$\sum_{n=1}^{\infty} \alpha^{1+\varepsilon}(n) < \infty$$

but

$$\sigma_n^{-1} S_n \not\to N(0,1), \qquad\qquad n \to \infty.$$

In connection with Davydov's examples it is natural to ask if
it is possible to weaken the requirements of Theorem 5.1.4 pro-
viding that for the process the next relation is valid

$$\sigma_n^2 \sim \sigma_0^2 n, \qquad\qquad n \to \infty, \; \sigma^2 > 0. \qquad (5.1.9)$$

As it shown Herrndorf's examples (see Chapter 2) in the class
of s.r.p. ξ_t, $t \in \mathbb{Z}^1$, such that $E\xi_0^2 < \infty$ and $\sigma_n^2 \sim \sigma^2 n$
as $n \to \infty$ none of a mixing rate guarantees the realizability
of the c.l.th. Recently, developing Herrndorf's idea Bradley
/14/ has shown the optimality of the conditions of Theorem
5.1.4 under the additional assumption (5.1.9).

<u>Theorem 5.1.10 (/14/)</u>. There exists an orthogonal β-mixing
s.r.p. ξ_t, $t \in \mathbb{Z}^1$, such that $E\xi_0 = 0$, $E|\xi_0|^{2+\delta} < \infty$,
$\delta > 0$, $\beta(n) = O(n^{-1}\log^3 n)^{\frac{2+\delta}{\delta}}$ as $n \to \infty$ and

$$\inf_{n \geq 1} P(S_n = 0) \cdot 0 \;, \; \lim_{C \to \infty} \left[\sup_{n \geq 1} P(|S_n| > C) \right] = 0. \qquad (5.1.10)$$

<u>Theorem 5.1.11 (/14/)</u>. There exists an orthogonal β-mixing
s.r.p. ξ_t, $t \in \mathbb{Z}^1$, such that $E\xi_0 = 0$, $|\xi_0| < C$ a.s.,
for some $C < \infty$, $\beta(n) = O\left(\frac{(\log n)^3}{n}\right)$ as $n \to \infty$ and

$$\inf_{n \geq 1} P(S_n = 0) > 0 \;, \; \lim_{C \to \infty} \left[\sup_{n \geq 1} P(|S_n| > C) \right] = 0.$$

Another question connected with Theorem 5.1.4 is the following:if
it possible to prove the c.l.th. in the case $\sigma^2 = 0$ and
$\sigma_n^2 = n h(n)$, where $h(n)$ is a slowly varying function.
The answer (see Bradley /14/) is such:

<u>Theorem 5.1.12</u>. There exists a β-mixing, centered s.r.p.
ξ_t, $t \in \mathbb{Z}^1$, such that $E|\xi_0|^{2+\delta} < \infty$, $\delta > 0$, $\sum_{n=1}^{\infty}(\beta(n))^{\delta/2+\delta} < \infty$

$$DS_n \sim (\log n)^{-4} n , \qquad n \to \infty$$

and (5.1.10) holds.

Theorem 5.1.14. There exists a β -mixing, centered s.r.p. ξ_t, $t \in \mathbb{Z}^1$, such that $|\xi_0| < C$, a.s. for some $C < \infty$

$$\sum_{n=1}^{\infty} \beta(n) < \infty, \qquad DS_n \sim (\log n)^{-4} n , \qquad n \to \infty$$

and (5.1.10) holds.

Clearly that β -mixing condition (see Chapter 2) for a s.r.p. implies α -mixing one.

Let us mention another result of Bradley, which is the consequence of Theorem 5.1.4 and Lemma 3.1.8.

Theorem 5.1.15. Let ξ_t, $t \in \mathbb{Z}^1$ be a s.r.p., centered, $n \to \infty$ and one of the following two conditions hold:

1) for some $\delta > 0$, $E|\xi_0|^{2+\delta} < \infty$ and $\sum_{n=1}^{\infty} n(\alpha(n))^{\frac{\delta}{2+\delta}} < \infty$

or

2) for some $C > 0$, $|\xi_0| < C$ a.s. and $\sum_{n=1}^{\infty} n \alpha(n) < \infty$.

Then the c.l.th. for ξ_t, $t \in \mathbb{Z}^1$, holds.

Bradley /14/ also showed that these conditions indicate the slowest mixing rates for $\alpha(n)$ that will guarantee the c.l.th. without the additional requirement $\sigma_0^2 > 0$.

5.2 Rate of Convergence in Central Limit Theorem for

α -mixing Stationary Random Processes

In this section our aim is not to prove the best of the existing estimates of the convergence rate in the c.l.th. Here we will give a simple method for obtaining estimates of the convergence rate which is quite sufficient to establish, for example, the law of the iterated logarithm. Simultaneously, the proof of the c.l.th. will be obtained by Bernstein's method, without omitting the small blocks.

Theorem 5.2.1. Suppose that a centered s.r.p. ξ_t, $t \in \mathbb{Z}^1$, satisfies the α -mixing condition and for some $\delta > 0$ $E|\xi_0|^{2+\delta} < \infty$. If in addition

$$\sum_{n\geq\tau} \alpha^{\delta/2+\delta}(n) = O\left(\frac{1}{\ln \tau}\right)^{3+\omega}, \quad \tau\to\infty, \quad \omega > 0 ,$$

then $\quad \sigma^2 = E\,\xi_0^2 + 2\sum_{j=1}^{\infty} E\,\xi_0\,\xi_j < \infty \qquad$ and if $\quad \sigma^2 > 0$

therefore there exists a real number $\quad \rho > 0 \quad$, such that

$$\Delta_n = \sup_{-\infty < x < \infty} \left| P\left(\frac{S_n}{\sigma_n} < x\right) - \Phi(x)\right| \leq \frac{C}{\ln^{1+\rho} n} , \quad n \in \mathbb{N} .$$

<u>Proof.</u>   Denote

$$S_{pj} = \frac{1}{\sigma_n}\sum_{t=p(j-1)}^{pj}\xi_t , \quad j=\overline{1,k-1}, \quad S_{pk}=\frac{1}{\sigma_n}\sum_{t=pk}^{n}\xi_t ,$$

$$k = \left[\frac{n}{p}\right].$$

Then

$$\left| P\left(\frac{S_n}{\sigma_n} < x\right) - \Phi(x)\right| \leq \left| P\left(\sum_{j=1}^{K} S_{pj} < x\right) - P\left(\sum_{j=1}^{K}\widetilde{S}_{pj} < x\right)\right| +$$

$$+ \left| P\left(\sum_{j=1}^{k}\widetilde{S}_{pj} < x\right) - \Phi(x)\right| , \tag{5.2.1}$$

where the random variables $\quad \widetilde{S}_{pj}, \ j=\overline{1,k} \quad$ are independent
and such that $\quad P(S_{p,j} < x) = P(\widetilde{S}_{pj} < x), \quad j=\overline{1,k}, \quad x\in R^1.$

Using inequality $(3.3.7)$ and Lemma 2.1.1 we obtain

$$\left| E\prod_{j=1}^{K} e^{it\,S_{pj}} - \prod_{j=1}^{K} E\,e^{it\,S_{pj}}\right| \leq$$

$$\leq \sum_{j=1}^{k}\left| E\left(e^{it\,S_{pj}}-1\right)\left(e^{it\,S_{pj+1}}-1\right)\prod_{m=j+2}^{K} e^{it\,S_{pm}} -\right.$$

$$\left. - E\left(e^{it\,S_{pj}}-1\right)E\left(e^{it\,S_{pj+1}}-1\right)\prod_{m=j+2}^{K} e^{it\,S_{pm}}\right| +$$

$$+ \sum_{j=1}^{k-1}\sum_{m=j+2}^{k}\left| E\left(e^{it\,S_{pj}}-1\right)\left(e^{it\,S_{pm}}-1\right)\prod_{s=m+1}^{K} e^{it\,S_{ps}} -\right.$$

$$\left. - E\left(e^{it\,S_{pj}}-1\right)E\left(e^{it\,S_{pm}}-1\right)\prod_{s=m+1}^{K} e^{it\,S_{ps}}\right| \leq$$

$$\leq \sum_{j=1}^{k} \left| E(e^{itS_{Pj}} - itS_{Pj} - 1)(e^{itS_{Pj+1}} - 1) \prod_{m=j+2}^{k} e^{itS_{Pm}} \right| +$$

$$+ |t| \sum_{j=1}^{k} \left| ES_{Pj}(e^{itS_{Pj+1}} - itS_{Pj+1} - 1) \prod_{m=j+2}^{k} e^{itS_{Pm}} \right| +$$

$$+ t^2 \sum_{j=1}^{k} |ES_{Pj}S_{Pj+1}| + \sum_{j=1}^{k} \left| E(e^{itS_{Pj}} - itS_{Pj} - 1) \times \right.$$

$$\times \left. E(e^{itS_{Pj+1}} - 1) \prod_{m=j+2}^{k} e^{itS_{Pm}} \right| +$$

$$+ \sum_{j=1}^{k} \sum_{m=j+2}^{k} E^{\frac{1}{2+\delta}} |e^{itS_{Pj}} - 1|^{2+\delta} E^{\frac{1}{2+\delta}} |e^{itS_{Pm}} - 1|^{2+\delta} \alpha^{\frac{\delta}{2+\delta}}(|m-j-1|p).$$

Using Lemma 2.1.1 again, inequality 4.2.4 and Hölder's inequality we continue the previous relations:

$$\left| E \prod_{j=1}^{k} e^{itS_{Pj}} - \prod_{j=1}^{k} E e^{itS_{Pj}} \right| \leq$$

$$\leq 3|t|^{2+\delta} k E \left| \frac{S_P}{6n} \right|^{2+\delta} + t^2 \sum_{j=1}^{k} |ES_{Pj}S_{Pj+1}| +$$

$$+ t^2 E^{\frac{2}{2+\delta}} \left| \frac{S_P}{6n} \right|^{2+\delta} \sum_{j=1}^{k} \sum_{m=j+2}^{k} \alpha^{\frac{\delta}{2+\delta}}(|m-j-1|p) \leq$$

$$\leq 3|t|^{2+\delta} k E \left| \frac{S_P}{6n} \right|^{2+\delta} + \frac{Ct^2}{P} \sum_{j=1}^{P} \sum_{k \geq j}^{\infty} \alpha^{\frac{\delta}{2+\delta}}(k) +$$

$$+ t^2 E^{\frac{2}{2+\delta}} \left| \frac{S_P}{6n} \right|^{2+\delta} k \sum_{j=1}^{k} \alpha^{\frac{\delta}{2+\delta}}(jp),$$

$$0 < C < \infty.$$

By virtue of both Minkowski's inequality and (5.1.9) and taking into account that $k \sim \frac{n}{P}$, $n \to \infty$ we have

$$\left| E \prod_{j=1}^{K} e^{itS_{Pj}} - \prod_{j=1}^{K} E e^{itS_{Pj}} \right| \le C_1 |t|^{2+\delta} \frac{P^{1+\delta}}{n^{\delta/2}} +$$

$$\frac{C}{P} t^2 \sum_{j=1}^{P} \sum_{k \ge j}^{\infty} \alpha^{\frac{\delta}{2+\delta}}(k) + C_2 t^2 \sum_{j \ge P}^{\infty} \alpha^{\frac{\delta}{2+\delta}}(j) \le$$

$$\le C_1 |t|^{2+\delta} \frac{P^{1+\delta}}{n^{\delta/2}} + C_3 t^2 \frac{N}{P} + t^2 \sum_{k \ge N}^{\infty} \alpha^{\frac{\delta}{2+\delta}}(k) +$$

$$+ C_2 t^2 \sum_{j \ge P}^{\infty} \alpha^{\frac{\delta}{2+\delta}}(j) .$$

Now we will use the Berry-Esseen theorem. For the smoothness condition in this theorem to be satisfied one can act in the following way. Let ξ_n be a normally distributed random variable with sufficiently small variance, for instance,

$$\xi_n \sim N\left(0, \tfrac{1}{\sqrt{n}}\right).$$

Let ξ_n and $\dfrac{S_n}{\sigma_n}$ be independent. Now if we add ξ_n to $\sigma_n^{-1} S_n$ hence the distribution function of random variable $\sigma_n^{-1} S_n + \xi_n$ will have a finite density, but such correction will not influence the convergent rate. Thus, applying Theorems 1.6.1 and 1.6.2 we get

$$\left| P\left(\frac{S_n}{\sigma_n} < x \right) - \Phi(x) \right| \le$$

$$\le C_1' |T|^{2+\delta} \frac{P^{1+\delta}}{n^{\delta/2}} + \frac{C_5}{T} + C_3' T^2 \frac{N}{P} + C_4 T^2 \sum_{k \ge N} \alpha^{\frac{\delta}{2+\delta}}(k) + C_1 |T|^{2+\delta} \frac{P^{1+\delta}}{n^{\delta/2}} +$$

$$+ C_2' T^2 \sum_{j \ge P}^{\infty} \alpha^{\frac{\delta/2+\delta}{}}(j) .$$

Putting
$$P = n^{\gamma}, \quad \text{where} \quad 0 < \gamma < \frac{\delta}{2(1+\delta)}, \quad T = \ln^{1+\varepsilon} n, \quad N = n^{\beta}, \quad 0 < \beta < \gamma, \varepsilon < \frac{\omega}{2}$$
we obtain our statement.

Similarly the following theorem is proved.

<u>Theorem 5.2.2.</u>   Let ξ_t, $t \in \mathbb{Z}'$, be a centered, α -mixing s.r.p. such that with probability 1 $|\xi_0| < C$, $0 < C < \infty$, and

$$\sum_{n \geq \tau} \alpha(n) = O\left(\frac{1}{\ln \tau}\right)^{3+\omega}, \qquad \omega > 0 .$$

Then $\quad \sigma_0^2 = E \xi_0^2 + 2 \sum_{j=1}^{\infty} E \xi_0 \xi_j < \infty \qquad$. If in addition $\sigma_0^2 > 0$, then there exists a real number $\rho, \rho > \omega$, such that

$$\Delta_n \leq \frac{C}{\ln^{1+\rho} n} , \qquad n \in \mathbb{N} .$$

In conclusion we will give without proof some results of Tihomirov /123/ concerning the rate of convergence in c.l.th. for α -mixing s.r.p.

<u>Theorem 5.2.3.</u>   Suppose a s.r.p. ξ_t , $t \in \mathbb{Z}^1$, satisfies α -mixing condition and there exist constants $K > 0$ and $\beta > 1$ such that for all $n \in \mathbb{N}$ and some δ , $0 < \delta \leq 1$,

$$E |\xi_0|^{2+\delta} < \infty \qquad \text{and} \qquad \alpha(n) \leq K n^{-\frac{\beta(2+\delta)(1+\delta)}{\delta^2}} .$$

Then if $\quad \sigma_0^2 = E \xi_0^2 + 2 \sum_{j=1}^{\infty} E \xi_0 \xi_j \neq 0 \qquad$ we have

$$\Delta_n \leq C_0 n^{-\frac{\delta}{2} \frac{\beta-1}{\beta+1}} .$$

But if $\quad \alpha(n) \leq C_1 e^{-C_2 n}$, $\quad C_1, C_2 > 0$, $\quad n \in \mathbb{N}$.

Then

$$\Delta_n \leq C_3 n^{-\frac{\delta}{2}} \ln^{1+\delta} n .$$

5.3 Invariance Principle for Stationary Random Processes
with α -mixing Condition

The first results concerning the invariance principle for α-mixing s.r.p. were obtained by Davydov /24/ . Below we will represent the results of Oodaira and Ioshihara /97/ who show the validity of the invariance principle under the condition of Theorem 5.1.7.

<u>Theorem 5.3.1.</u>   Let ξ_t , $t \in \mathbb{Z}^1$, be a centered α -mixing s.r.p., such that one of the following two conditions holds:

1) $|\xi_t| < C$ with probability 1 and $\sum_{n=1}^{\infty} \alpha(n) < \infty$;

2) $E |\xi_t|^{2+\delta} < \infty$ for some $\delta > 0$ and $\sum_{n=1}^{\infty} \alpha^{\frac{\delta}{2+\delta}}(n) < \infty$.

Then the series $\quad \sigma^2 = E \xi_0^2 + 2 \sum_{j=1}^{\infty} E \xi_0 \xi_j \qquad$ converges. If in addition $\delta^2 > 0$, then the following sequence of random processes

$$X_n(t) = \frac{S_{k-1}}{\delta_n} + \frac{t - \frac{k-1}{n}}{\frac{1}{n}} \frac{\xi_k}{\delta_n} , \qquad t \in \left[\frac{k-1}{n}, \frac{k}{n} \right], \quad k = \overline{1,n}$$

converges weakly to Wiener process W .

<u>Proof.</u> According to Prohorov's theorem (see Chapter 1) it is sufficient to check that the finite-dimensional distributions $P_n(t_1, \ldots, t_k)$ of the processes $X_n(t)$ converge weakly to those of Wiener process and that the sequence $X_n(t)$, $n \in \mathbb{N}$ is tight. First let's prove two technical lemmas which will be useful later on.

<u>Lemma 5.3.1.</u> Let ξ_t , $t \in \mathbb{Z}^1$, be a centered α -mixing s.r.p. such that the next conditions are valid:

1. $\sigma_n^2 \to \infty$, $n \to \infty$

2. $\sigma_n^{-1} S_n \xrightarrow[n \to \infty]{\mathcal{D}} N(0,1)$.

The the finite-dimensional distributions $P_n(t_1, \ldots, t_k)$ of the processes $X_n(t)$ converge weakly to those $W_n(t_1, \ldots, t_k)$ of Wiener process.

<u>Proof.</u>   Let us fix the points $t_1 < t_2 < \ldots < t_k$, $t_j \in [0,1]$, $j = \overline{1,k}$, and consider the random variable

$$Z_n = \sum_{j=1}^{k} \tau_j X_n(t_j) = \sum_{j=1}^{k} a_j \left(X_n(t_j) - X_n(t_j - 1) \right),$$

$$t_0 = 0 , \qquad a_j = \sum_{i=j}^{k} \tau_j .$$

Let $f_n(\tau)$ be its characteristic function.

Suppose that $g(n)$, $n \in \mathbb{N}$, is a sequence of natural numbers, such that

$$g(n) \sigma_n^{-2} \to 0 , \qquad n \to \infty$$

and the numbers m_j , $j = \overline{1,k}$ are defined from the condition

$$m_j n^{-1} \leq t_j < m_{j+1} n^{-1}.$$

Put $p(j) = m_j - g(j)$ for $1 \leq j < k$, $p(k) = m(k)$ and

$$S_k^m = \sum_{j=k}^{m-1} \xi_j .$$

Then we can write

$$X_n(t_j) - X_n(t_{j-1}) = \frac{1}{\sigma_n} S_{m_j} + \frac{t_j - \frac{m_j}{n}}{\frac{1}{n}} \cdot \frac{1}{\sigma_n} \xi_{m_{j+1}} -$$

$$- \frac{1}{\sigma_n} S_{m_{j-1}} - \frac{t_{j-1} - \frac{m_{j-1}}{n}}{\frac{1}{n}} \cdot \frac{1}{\sigma_n} \xi_{m_j} =$$

$$= \frac{1}{\sigma_n} S_{m_{j-1}}^{m_j} + \frac{t_j - \frac{m_j}{n}}{\frac{1}{n}} \cdot \frac{1}{\sigma_n} \xi_{m_{j+1}} - \frac{t_{j-1} - \frac{m_{j-1}}{n}}{\frac{1}{n}} \cdot \frac{1}{\sigma_n} \xi_{m_j} =$$

$$= \frac{1}{\sigma_n} S_{m_{j-1}}^{P(j)} + T_j^{(n)},$$

$$T_j^{(n)} = \frac{1}{\sigma_n} S_{P(j)}^{m_j} + \frac{t_j - \frac{m_j}{n}}{\frac{1}{n}} \cdot \frac{1}{\sigma_n} \xi_{m_{j+1}} - \frac{t_{j-1} - \frac{m_{j-1}}{n}}{\frac{1}{n}} \cdot \frac{1}{\sigma_n} \xi_{m_j}.$$

Hence

$$\sum_{j=1}^{k} a_j (X_n(t_j) - X_n(t_{j-1})) = \sum_{j=1}^{k} \frac{a_j}{\sigma_n} S_{m_{j-1}}^{P(j)} + \sum_{j=1}^{k} a_j T_j^{(n)}. \qquad (5.3.1)$$

Let us show that under the conditions of Lemma 5.3.1

$$D\left(\sum_{j=1}^{k} a_j T_j^{(n)} \right) \to 0, \quad n \to \infty$$

and hence
the sum $\sum_{j=1}^{k} a_j T_j^{(n)}$ does not influence the limit distribution of (5.3.1). So it is sufficient to show that

$$\frac{D S_{P(j)}^{m_j}}{\sigma_n^2} \to 0, \qquad \left(t_j - \frac{m_j}{n} \right)^2 D \xi_{m_{j+1}} \to 0, \quad n \to \infty$$

for any $j = \overline{1, k}$.

Note that according to Remark 3.1.4 under the conditions of Lemma 5.2.1 we have

$$\sigma_n^2 = n h(n),$$

where $h(n)$ is a slowly varying function.
We can write

$$\frac{D S_{P(j)}^{m_j}}{\sigma_n^2} = \frac{h(g(n)) g(n)}{n h(n)} = \frac{g(n)}{n} \frac{h\left(n \frac{g(n)}{n} \right)}{h(n)} \leq$$

$$\leq C \frac{g(n)}{n} \left(\frac{n}{g(n)}\right)^{\tau} \to 0, \quad n \to \infty \quad (0 < \tau < 1).$$

Here we use Property 4 of a slowly varying function (see Chapter 1).

Further

$$\left(t_j - \frac{m_j}{n}\right)^2 D\xi_{m_{j+1}} \leq \frac{C}{n^2} \to 0, \quad n \to \infty.$$

By virtue of the stationary state of the considered process the random variables $Z_n^j = a_j S_{m_{j-1}}^{P_j} \sigma_n^{-1}$ have the identical distribution which coincides with that of the following random variable:

$$a_j S_{P_j - m_{j-1}} \sigma_n^{-1} = \frac{a_j S_{P_j - m_{j-1}}}{\sigma_{P_j - m_{j-1}}} \cdot \frac{\sigma_{P_j - m_{j-1}}}{\sigma_n}.$$

By the conditions of Lemma 5.3.1 the latter is distributed asymptotically normal with zero mean and variance

$$a_j^2 (t_j - t_{j-1}).$$

Using the inequality (3.3.5) we have

$$\left| E \prod_{j=1}^{K} e^{itZ_n^j} - \prod_{j=1}^{K} E e^{itZ_n^j} \right| \leq K\alpha(g(n))$$

and since $\alpha(g(n)) \to 0$, $n \to \infty$ it follows that

$$E e^{i\tau Z_n^j} \longrightarrow \exp\left\{-\tfrac{1}{2} a_j^2 (t_j - t_{j-1})\tau^2\right\}, \quad n \to \infty.$$

Hence

$$E e^{i\tau \sum_{j=1}^{K} Z_n^j} \longrightarrow e^{-\sum_{j=1}^{K} a_j^2 (t_{j-1} - t_{j-1})\frac{\tau^2}{2}}.$$

The same limit has also $f_n(\tau)$ as $n \to \infty$. In particular,

$$f_n(1) \xrightarrow[n \to \infty]{} e^{-\frac{1}{2} \sum_{j=1}^{K} a_j^2 (t_j - t_{j-1})}.$$

Since $f_n(1)$ is the characteristic function of the vector $(X_n(t_1), \ldots, X_n(t_K))$ and correspondingly $\exp\left\{-\tfrac{1}{2} \sum_{j=1}^{K} a_j^2 (t_j - t_{j-1})\right\}$ is the characteristic function of $(W(t_1), \ldots, W(t_K))$ that's why we finish the proof.

Corollary 5.3.1. Suppose that ξ_t, $t \in \mathbb{Z}^1$, is a centered α-mixing s.r.p., $E\xi_0^2 < \infty$, the sequence $(S_n \sigma_n^{-1})^2$ is uniformly integrable and $\sigma_n^2 \to \infty$ as $n \to \infty$. Then for the process ξ_t, $t \in \mathbb{Z}^1$ the invariance principle is fulfilled iff the following condition is valid:

for any $\varepsilon > 0$ there exists a λ, $\lambda > 1$ such that

$$P\left(\max_{1 \leq i \leq n} |S_i| > \lambda \sigma_n\right) \leq \lambda^{-2} \varepsilon. \qquad (5.3.2)$$

__Corollary 5.3.2.__ Let ξ_t, $t \in \mathbb{Z}^1$ be a centered α -mixing s.r.p. such that $E \xi_0^2 < \infty$ and $\sigma_n^2 \to \infty$ as $n \to \infty$. If for some $\delta > 0$

$$E |S_n|^{2+\delta} < C \sigma_n^{2+\delta}, \qquad 0 < C < \infty, \qquad (5.3.3)$$

then for the process the invariance principle holds.

__Proof.__ Indeed, it is sufficient to show that the conditions of the previous corollary are valid. Relation (5.3.3) implies the uniform integrability of the sequence $(S_n \sigma_n^{-1})^2$. To check the validity of (5.3.3) we will use the next theorem.

__Theorem(Billingsley /7/).__ Let

$$P(|S_j - S_i| \geqslant \lambda) \leq \lambda^{-\gamma} \left(\sum_{i < \ell \leq j} u_\ell\right)^\alpha, \qquad 0 \leq i \leq j \leq n,$$

$$u_\ell \geqslant 0, \qquad \alpha > 1, \qquad \gamma \geqslant 0 \qquad \qquad . \text{ Then}$$

$$P\left(\max_{1 \leq j \leq n} |S_j| \geqslant \lambda\right) \leq \lambda^{-\gamma} C_{\gamma, \alpha} (u_1 + \cdots + u_n)^\alpha.$$

Taking into account (5.3.3) and Property 4 of slowly varying functions we can write

$$P(|S_j - S_i| \geqslant \lambda \sigma_n) \leq \lambda^{-(2+\delta)} \sigma_n^{-(2+\delta)} E|S_j - S_i|^{2+\delta} \leq$$

$$\leq C_1 \lambda^{-(2+\delta)} \left(\frac{(j-i) h(j-i)}{n h(n)}\right)^{\frac{2+\delta}{2}} \leq C_1 \lambda^{-(2+\delta)} \left(\sum_{\ell=i}^{j} \frac{1}{n}\right)^{\frac{\tau(2+\delta)}{2}}, \qquad 0 < \tau < 1.$$

Hence

$$P\left(\max_{1 \leq i \leq n} |S_i| \geqslant \lambda \sigma_n\right) \leq \frac{C_3}{\lambda^{2+\delta}} .$$

Corollary 5.3.2 is proven.

__Lemma 5.3.2.__ Under conditions of Theorem 5.3.1 the sequence $X_n(t)$ $n \in \mathbb{N}$ is tight.

__Proof.__ By Theorem 1.3.3 for the sequence $X_n(t)$, $n \in \mathbb{N}$ to be tight it is sufficient to show that for any $\varepsilon > 0$ there exist

λ, $\lambda > 1$ and an integer n_0 , such that

$$P\left(\sigma_n^{-1} \max_{1 \le i \le n} |S_i| \ge \lambda\right) \le \frac{\varepsilon}{\lambda^2} \ , \qquad n > n_0 \ . \tag{5.3.4}$$

Using inequality (3.3.1) we can write

$$P\left(\max_{1 \le i \le n} |S_i| \ge \lambda \sigma_n\right) \le$$

$$\le \frac{P\left(|S_n| \ge \frac{\lambda \sigma_n}{2}\right) + k\alpha(p) + 2\frac{n}{p} P\left(\sum_{i=1}^{2p} |\xi_i| \ge \frac{\lambda \sigma_n}{2}\right)}{1 - \max_{1 \le i \le k-2} P\left(|S_n - S_{(i+2)p}| \ge \frac{\lambda \sigma_n}{2}\right)} \tag{5.3.5}$$

Let $P = [n^{\delta/2(1+\delta)}]$, $\left(P = 0(\sqrt{n})\right)$.

Then $k\alpha(p) \to 0$, $n \to \infty$ \tag{5.3.6}

By Chebyshev's inequality

$$\max_{1 \le i \le k-2} P\left(|S_n - S_{(i+2)p}| \ge \frac{\lambda \sigma_n}{2}\right) \le$$

$$\le \max_{1 \le i \le n} P\left(|S_i| \ge \frac{\lambda \sigma_n}{2}\right) \le \frac{4}{\lambda^2} \max_{1 \le i \le n} \frac{i\, h(i)}{n\, h(n)} \le \frac{4}{\lambda^2}$$

Then for sufficiently large λ and all $n \in \mathbb{N}$ we have

$$\max_{1 \le i \le k-2} P\left(|S_n - S_{(i+2)p}| \ge \frac{\lambda \sigma_n}{2}\right) \le \frac{1}{2} \ . \tag{5.3.7}$$

By virtue of uniform integrability of sequence $\sigma_n^{-2} S_n^2$, $n \in \mathbb{N}$ the following relation is valid for any $\varepsilon' > 0$ and sufficiently large n :

$$P\left(|S_n| \ge \frac{\lambda \sigma_n}{2}\right) \le \frac{4}{\lambda^2} E\left|\frac{S_n}{\sigma_n}\right|^2 I\left(\left|\frac{S_n}{\sigma_n}\right| \ge \lambda\right) \le \frac{\varepsilon'}{\lambda^2} \ , \quad n \in \mathbb{N} \ . \tag{5.3.8}$$

Now using Chebyshev's inequality and farther Minkovski's inequality we obtain for sufficiently large

$$\frac{n}{p} P\left(\sum_{i=1}^{2p} |\xi_i| \ge \frac{\lambda \sigma_n}{2}\right) \le C_1 \frac{n}{p} \cdot \frac{p^{2+\delta}}{\lambda^{2+\delta} \sigma_n^{2+\delta}} \le$$

$$\leq \frac{C_2}{\lambda^{2+\delta}} \; \frac{n \cdot n^{\frac{\delta(2+\delta)}{2(1+\delta)}}}{n^{\frac{\delta}{2(1+\delta)}} \, n^{1+\delta/2}} = \frac{C_2}{\lambda^{2+\delta}} \; . \qquad\qquad (5.3.9)$$

Substituting (5.3.6) - (5.3.9) into (5.3.5) we obtain (5.3.4). Lemma 5.3.2 is proven.

Now it is clear that for proving Theorem 5.3.1 it is sufficient to show that the conditions of Lemmas 5.3.1 and 5.3.2 are fulfilled. Since the conditions of Theorem 5.3.1 are the same as those of Theorem 5.3.4 we have

$$\sigma_n^2 \sim \sigma_0^2 \, n, \qquad \sigma_n^{-1} S_n \xrightarrow{\mathcal{D}} N(0,1), \qquad n \to \infty \;.$$

Thus the conditions of Lemma 5.3.1 are valid and hence the finite-dimensional distributions $P_n(t_1, \ldots, t_k)$ converge weakly to those of Wiener process W. According to Theorem 5.1.2 the sequence $(\sigma_n^{-1} S_n)^2, \; n \in \mathbb{N}$ is uniformly integrable and it follows from conditions of Theorem 5.3.1 that the sequence

$$X_n(t), \; n \in \mathbb{N}$$

is tight.

Recently Herrndorf /59/ has generalized these results to the non-stationary case and also weaken the restrictions for the moments. Namely, he obtained the theorems formulated below.
Let $\xi_t, \; t \in \mathbb{Z}^1$ be a centered random sequence having finite second moments and such that:

$$n^{-1} E S_n^2 \to \sigma^2 > 0, \qquad n \to \infty, \qquad\qquad (5.3.10)$$

$$\sup \left\{ E(S_{m+n} - S_m)^2 n^{-1}, \qquad n, m \in \mathbb{N} \right\} \qquad (5.3.11)$$

and let

$$\alpha(k) = \sup_{n \in \mathbb{N}} \alpha_n(k), \quad k \in \mathbb{N},$$

$$\alpha_n(k) = \sup \left\{ |P(AB) - P(A)P(B)|; \quad A \in \mathcal{O}_1^m, \quad B \in \mathcal{O}_{m+k}^n, \right.$$

$$\left. 1 \leq m \leq n-k \right\}, \qquad k \leq n-1,$$

$$\alpha_n(k) = 0, \qquad k \geq n \;.$$

Denote

$$\|X\|_\beta = E^{\frac{1}{\beta}} |X|^\beta, \qquad \beta \in [1, +\infty),$$

$$\| X \|_\beta = \text{ess sup } |X|, \quad \beta = \infty \, .$$

Theorem 5.3.2. Let $\beta \in (2, \infty]$ and $\gamma = \frac{2}{\beta}$ $\left(\frac{2}{\infty} = 0\right)$.

Let a_n, $n \in \mathbb{N}$ be a number sequence from $[1, \infty)$, If a r.p. ξ_t, $t \in \mathbb{N}$ satisfies the conditions (5.3.10) , (5.3.11) and

$$\left(\sup_{1 \leq i \leq n} \| X_i \|_\beta^2 \right) \left(\sum_{j \geq a_n} \alpha_n^{1-\gamma}(i) + \frac{a_n^{2-\gamma}}{n^{1-\gamma}} \right) \to 0, \quad n \to \infty,$$

then the sequence $X_n(t)$, $n \in \mathbb{N}$ converges weakly to Wiener process W .

Corollary 5.3.3 Let $\beta \in (2, \infty)$ and $\gamma = \frac{2}{\beta}$. If a r.p. ξ_t, $t \in \mathbb{N}$, satisfies the conditions (5.3.10), (5.3.11) and

$$\sum_{i=1}^{\infty} \alpha^{1-\gamma}(i) < \infty, \qquad \lim_{n \to \infty} \sup \| X_n \|_\beta < \infty,$$

then $X_n(t)$, $n \in \mathbb{N}$, converges weakly to W .

5.4 Law of Iterated Logarithm for Stationary Random

Processes Satisfying the α-mixing Condition

It is said that a s.r.p. ξ_t, $t \in \mathbb{Z}^1$, satisfies the law of iterated logarithm if

$$P \left(\lim_{n \to \infty} \sup \frac{S_n}{\sigma \sqrt{2n \log \log n}} = 1 \right) = 1, \quad \sigma > 0 \, .$$

Theorem 5.4.1 (/96/ , see also /109/). Suppose that a centered α-mixing s.r.p. ξ_t, $t \in \mathbb{Z}^1$, is such that

$$E \, \xi_0^2 < \infty, \quad \sum_{n=1}^{\infty} \alpha(n) < \infty \qquad \text{and}$$

1) $\qquad \sigma_n^2 = n \sigma^2 (1 + 0(1)), \qquad n \to \infty, \ \sigma^2 > 0.$

Then the process ξ_t, $t \in \mathbb{Z}^1$, satisfies the law of iterated logarithm if for any $\rho > 0$ and for all sufficiently large n the relations

2) $\qquad \sup_{x \in R^1} |P(S_n < x \sigma \sqrt{n}) - \Phi(x)| = O\left(\frac{1}{(\log n)^{1+\rho}} \right) \, ;$

3) $\qquad P(\max_{1 \leq j \leq n} |S_j| \geq \theta \chi(n)) = O\left(\frac{1}{(\log n)^{1+\rho}} \right) \, ;$

$\qquad \theta > 1, \qquad \chi(n) = (2\sigma^2 n \log \log n)^{\frac{1}{2}}$

are valid.

<u>Proof.</u> It is sufficient to show that for any $\varepsilon > 0$

$$P\left(|S_n| > (1+\varepsilon)\,\chi(n),\ \ i.o.\right) = 0 \qquad\qquad (5.4.1)$$

and

$$P\left(|S_n| > (1-\varepsilon)\,\chi(n),\ \ i.o.\right) = 1. \qquad\qquad (5.4.2)$$

First let us prove relation (5.4.1). For an arbitrary positive number τ one can choose a non-decreasing sequence of natural numbers n_k, $k \in \mathbb{N}$, such that

$$(n_k - 1)\,\sigma^2 \le (1+\tau)^k < n_k\,\sigma^2$$

for all $k > k_0$. Hence

$$\sigma^2 n_k \asymp (1+\tau)^k, \quad k > k_0$$

and

$$n_k - n_{k-1} = n_k\left(1 - \frac{n_{k-1}}{n_k}\right) \asymp n_k\,\frac{\tau}{1+\tau}\ , \quad k > k_0.$$

From Condition 2) of Theorem 5.4.1 we obtain

$$P\left(\max_{1 \le j \le n_k} |S_j| > (1+\gamma)\,\chi(n_k)\right) \le$$

$$\le C\,(\log n_k)^{-(1+\rho)} < C\,(k\,\log(1+\tau))^{-(1+\gamma_1)}$$

for any $\gamma > 0$, $k > k_0$, and $0 < \gamma_1 < \rho$.
Hence

$$\sum_{k=1}^{\infty} P\left(\max_{1 \le j \le n_k} |S_j| > (1+\gamma)\,\chi(n_k)\right) < \infty.$$

Note that for all sufficiently large k

$$\frac{\chi(n_k)}{\chi(n_{k-1})} < \sqrt{1+2\tau}\ .$$

For fixed γ, $0 < \gamma < \varepsilon$ let us choose a positive constant such that

$$\frac{1+\varepsilon}{\sqrt{1+2\tau}} > 1+\gamma.$$

Then by means of the Borel-Cantelli lemma we get

$$P\left(|S_n| > (1+\varepsilon)\,\chi(n)\ \ i.o.\right) \le$$

$$\le P\left(\max_{n_{k-1} < n < n_k} |S_n| > (1+\varepsilon)\,\chi(n_{k-1})\ \ i.o.\right) \le$$

108

$$\leqslant P\left(\max_{1\leqslant n<n_K} |S_n| > (1+\varepsilon)\,\chi(n_{K-1})\ \ i.o.\right)\leqslant$$

$$\leqslant P\left(\max_{1\leqslant n<n_K} |S_n| > \frac{1+\varepsilon}{\sqrt{1+2\tau}}\,\chi(n_{K-1})\ \ i.o.\right)\leqslant$$

$$\leqslant P\left(\max_{1\leqslant n<n_K} |S_n| > (1+\gamma)\,\chi(n_{K-1})\ \ i.o.\right)=0.$$

It remains to prove the relation (5.4.2). For sufficiently large $p\in\mathbb{N}$ and sufficiently small $\delta>0$ and $\varepsilon>0$ let us introduce the events:

$$E_j = \left\{|S_{p^j}| \leqslant (1-\delta)\,\chi(p^j)\,,\quad j=1,\ldots,k-1\right\},$$

$$E_k = \left\{|S_{p^k}| > (1-\delta)\,\chi(p^k)\right\},$$

$$G_K = \left\{|S_{p^k} - S_{p^{k-1}+n_k}| > (1-\gamma)\chi(p^k)\right\},$$

$$L_K = \left\{|S_{p^{k-1}+n_K} - S_{p^{k-1}}| < \varepsilon\,\chi(p^k)\right\}.$$

Here $k\in\mathbb{N}$, $n_K=[p^{\frac{K}{2}}]$ and γ is chosen, such that the relation

$$\frac{2}{\sqrt{p}} + \gamma + \varepsilon < \delta$$

holds.

Put $u_k = P(E_k \cap F_k)$ and $\mathcal{U}_m = \sum_{k=1}^{m} u_k$.

From Condition 3) of Theorem 5.4.1 it follows that for all sufficiently large n and for any $\beta>1$ we have

$$P\left(|S_n| > \beta\,\chi(n)\right) > \frac{C_1}{(\log n)(\log\log n)},\quad C_1>0.$$

Since in addition for all sufficiently large p we have

$$m_K = p^k - (p^{k-1} + n_K) > \left[\frac{p^k}{2}\right],$$

then

$$\nu_K = P(G_K) \geqslant P\left(|S_{m_K}| > 2(1-\gamma)\chi(m_K)\right) \geqslant$$

$$\geq P\left(|S_{m_K}| > 2(1-\gamma)\chi(m_K)\right) > \frac{C_2}{k\log k} \; , \qquad C_2 > 0. \qquad (5.4.3)$$

The Chebyshev inequality implies that

$$P(L_K) \geq 1 - C_3 p^{-\frac{k}{2}} \; , \qquad C_3 > 0. \qquad (5.4.4)$$

Further we have

$$P(E_K \cap F_K) = u_K = \mathcal{U}_K - \mathcal{U}_{K-1} \geq P(E_K \cap G_K \cap L_K) \geq$$

$$\geq P(E_K \cap G_K) - P(\bar{L}_K) \geq P(E_K)P(G_K) - \alpha([p^{\frac{k}{2}}]) - P(\bar{L}_K).$$

From (5.4.4) and $\quad P(E_K) \geq 1 - \mathcal{U}_{K-1} \quad$ we obtain

$$\mathcal{U}_K \geq \mathcal{U}_{K-1} + v_K(1 - \mathcal{U}_{K-1}) - \alpha([p^{\frac{k}{2}}]) - C_3 p^{-\frac{k}{2}}.$$

Whence

$$1 - \mathcal{U}_K \leq (1 - v_K)(1 - \mathcal{U}_{K-1}) + \alpha([p^{\frac{k}{2}}]) + C_3 p^{-\frac{k}{2}}. \qquad (5.4.5)$$

From (5.4.3) it follows that the series $\displaystyle\sum_{K=1}^{\infty} v_K$ diverges and

$$\sum_{K=1}^{\infty} \left(\alpha([p^{\frac{k}{2}}]) + C_3 p^{-\frac{k}{2}}\right) \to 0, \qquad n \to \infty.$$

Then solving the recurrent relation (5.4.5) we get $\mathcal{U}_K \to 1$ as $k \to \infty$. Thus Theorem 5.4.1 is proven.

__Theorem 5.4.2.__ Let $\xi_t, t \in \mathbb{Z}^1$, be a α-mixing, centered s.r.p. and

1) $|\xi_0| < C$ with probability 1 ;

2) $\displaystyle\sum_{j \geq n} \alpha(j) = 0\left(\frac{1}{\ln^{3+\varepsilon} n}\right)$ for some $\varepsilon > 0$;

3) $\displaystyle\sigma_0^2 = E\xi_0^2 + 2\sum_{j=1}^{\infty} E\xi_0\xi_j > 0.$

Then the process $\xi_t, t \in \mathbb{Z}^1$, satisfies the law of iterated logarithm.

__Proof.__ It is sufficient to verify the validity of the conditions of Theorem 5.4.1. Condition 1) of Theorem 5.4.1. is evidently fulfilled, Condition 2) follows from Theorem 5.2.2. Let us verify Condition 3). For proving it let us use the inequality (3.3.1). We can write

$$P\left(\max_{1\le j\le n}|S_j|\ge \theta\chi(n)\right)\le \qquad\qquad (5.4.6)$$

$$\le \frac{P\left(|S_n|\ge \frac{\theta\chi(n)}{2}\right)+k\alpha(p)+2\frac{n}{p}P\left(\sum_{j=1}^{2p}|\xi_j|\ge \frac{1}{2}\theta\chi(n)\right)}{1-\max_{1\le i\le \kappa-2}P\left(|S_n-S_{(i+2)p}|\ge \frac{\theta\chi(n)}{2}\right)},$$

where

$$p=p(n)=\left[\ \sqrt{n}\ \right].$$

Further according to Condition 2) of Theorem 5.4.2 we have

$$\sum_{j\ge\sqrt{n}}^{n}\alpha(j)\le \frac{C}{\ln^{3+\varepsilon}n}.$$

Whence

$$(n-\sqrt{n})\alpha(n)\le \frac{C}{\ln^{3+\varepsilon}n}\qquad\text{and}\qquad \alpha(n)\le \frac{C_1}{n\ln^{3+\varepsilon}n}.$$

Finally

$$k\alpha(p)\sim \frac{n}{p}\alpha(p)\le \frac{C_2 n}{\sqrt{n}}\cdot\frac{1}{\sqrt{n}\ln^{3+\varepsilon}n}\le \frac{C_3}{\ln^{1+\rho}n},\quad \rho>0.\quad (5.4.7)$$

Now for sufficiently large n we can write

$$\max_{1\le i\le \kappa-2}P\left(|S_n-S_{(i+2)p}|\ge \theta\chi(n)\right)\le \qquad\qquad (5.4.8)$$

$$\le \max_{1\le i\le n}P\left(|S_i|\ge \theta\chi(n)\right)\le \frac{C_4}{\ln\ln n}\le \frac{1}{2}.$$

Using Chebyshev's and Minkowski's inequalities we obtain

$$P\left(\sum_{j=1}^{2p}|\xi_j|\ge \frac{1}{2}\theta\chi(n)\right)=0 \qquad\qquad (5.4.9)$$

Taking advantage the following property of Gaussian distribution (see, for example, /7/)

$$\int_{x}^{\infty}e^{-\frac{y^2}{2}}dy\sim \frac{1}{x}e^{-\frac{x^2}{2}},\quad x\to\infty$$

we have

$$P(|S_n| \geqslant \frac{\beta \chi(n)}{2}) = P(S_n \geqslant \frac{\beta \chi(n)}{2}) + P(S_n \leqslant -\frac{\beta \chi(n)}{2}) \leqslant$$

$$\leqslant |P((\sigma\sqrt{n})^{-1} S_n \geqslant \beta \sqrt{\log\log n}) + \Phi(\beta\sqrt{\log\log n}) - 1| +$$

$$+ |P((\sigma\sqrt{n})^{-1} S_n \leqslant -\beta\sqrt{\log\log n}) - \Phi(-\beta\sqrt{\log\log n})| +$$

$$+ 2\Phi(-\beta\sqrt{\log\log n}) \leqslant C_1 \Delta_p + C_2\, e^{-\beta^2 \log\log n} \leqslant$$

$$\leqslant C_1 (\log n)^{-(1+\rho)} + C_2 (\log n)^{-\beta^2}.$$

Putting $\quad \beta^2 = 1 + \rho \quad$ we obtain

$$P(|S_n| \geqslant 2^{-1}\beta\chi(n)) \leqslant C_3 (\log n)^{-(1+\rho)} \qquad (5.4.10)$$

Substituting (5.4.7)-(5.4.10) into (5.4.6) we finish the proof.

<u>Theorem 5.4.3.</u> A centered α-mixing s.r.p. ξ_t, $t \in \mathbb{Z}^1$ satisfies the law of iterated logarithm if the next conditions are fulfilled:

1) $\quad E|\xi_0|^{2+\delta} < \infty, \quad \delta > 0$

2) $\quad \sum\limits_{j \geqslant n} \alpha^{\delta/2+\delta}(j) = O(\frac{1}{\ln^{3+\varepsilon} n}), \quad n \to \infty, \quad \varepsilon > 0$

3) $\quad \sigma_0^2 = E\xi_0^2 + 2\sum\limits_{j=1}^{\infty} E\xi_0 \xi_j \neq 0$.

<u>Proof.</u> Put $\quad p = [n^{\delta/2(1+\delta)} \ln^{-1} n], \quad n \in \mathbb{N}$ .Than we have

$$\frac{n}{p}\alpha(p) \leqslant C_1 \frac{n \ln n}{n^{\delta/2(1+\delta)}} \cdot \frac{1}{p^{2+\delta/\delta}(\ln n)^{(3+\varepsilon)\cdot(\frac{2+\delta}{\delta})}} \leqslant C_1 \frac{1}{\ln^{1+\rho'} n}, \quad \rho' > 0$$

and

$$\frac{n}{p} P(\sum\limits_{j=1}^{2p} |\xi_j| \geqslant \frac{1}{2}\beta\chi(n)) \leqslant C_1 \frac{n}{pn^{2+\delta/\delta}} E(\sum\limits_{j=1}^{2p} |\xi_j|)^{2+\delta} \leqslant C_2 \frac{p^{1+\delta}}{n^{\delta/2}} = \frac{C_3}{\ln^{1+\delta} n} .$$

Further the proof continues as in the previous theorem.

<u>5.5 Limit Theorems for ρ -mixing Stationary Random Processes</u>

In /65/ Ibragimov established the following theorem.

<u>Theorem 5.5.1.</u> Let $\quad \xi_t$, $t \in \mathbb{Z}^1$ be a centered s.r.p. such that

$$E\xi_0^2 < \infty, \quad \sigma_n^2 \to \infty \quad \text{as} \quad n \to \infty \quad \text{and}$$

$$\sum_{j=1}^{\infty} \rho(2^j) < \infty.$$

Then ξ_t, $t \in \mathbb{Z}^1$, satisfies the c.l.th.

<u>Proof</u>. It is sufficient to check the conditions of Theorem 4.1.3. Let p, q be some standard pair of functions and let

$$S_n = \sum_{t=1}^{n} \xi_t \;, \qquad k = \left[\frac{n}{p+q}\right], \qquad X_j = \sum_{t=jp+jq+1}^{(j+1)p+jq} \xi_t \;,$$

$$\overline{X}_j = \sum_{t=(j+1)p+jq+1}^{(j+1)p+(j+1)q} \xi_t \;, \qquad j = 0, \ldots, k-1 \;,$$

$$X_k = \sum_{t=kp+kq+1}^{n} \xi_t \;, \qquad \overline{X}_k = 0 \;.$$

Clearly

$$S_n = \sum_{j=0}^{k} \left(X_j + \overline{X}_j \right) .$$

Let us represent $k = \left[\frac{n}{p+q}\right]$ in the following form

$$k = v_0 2^{\tau} + v_1 2^{\tau-1} + \ldots + v_{\tau-1} 2^1 + v_\tau \;, \quad v_j \in \{0,1\}, \quad j = \overline{1,\tau}, \quad v_0 = 1$$

and let

$$Y_0 = \frac{X_1}{\mathfrak{G}_n} \;, \quad Y_1 = \frac{X_2}{\mathfrak{G}_n} \;, \quad Y_j = \frac{1}{\mathfrak{G}_n} \sum_{t=2^{j-1}+1}^{2^j} X_t \;, \qquad j = \overline{2,\tau} \;,$$

$$\overline{Y}_0 = \frac{\overline{X}_1}{\mathfrak{G}_n} \;, \quad \overline{Y}_1 = \frac{\overline{X}_2}{\mathfrak{G}_n} \;, \quad \overline{Y}_j = \frac{1}{\mathfrak{G}_n} \sum_{t=2^{j-1}+1}^{2^j} \overline{X}_t \;, \qquad j = \overline{2,\tau} \;,$$

$$Z_0 = Z_1 = E\, e^{it\frac{X_1}{\mathfrak{G}_n}} \;, \qquad Z_j = \prod_{m=2^{j-1}+1}^{2^j} E\, e^{it\frac{X_m}{\mathfrak{G}_n}} \;, \qquad j = \overline{2,\tau} \;,$$

$$U_j = \left| E \prod_{s=0}^{j} e^{itY_s} - \prod_{s=0}^{j} Z_s \right| \;, \qquad j = \overline{0,\tau} \;,$$

$$\alpha_j = \left| E \prod_{s=0}^{j} e^{itY_s} - \prod_{s=0}^{j} E\, e^{itY_s} \right| \;, \qquad j = \overline{0,\tau} \;,$$

$$\hat{S}_j = \sum_{s=0}^{j} Y_s , \quad j = \overline{0, \tau} .$$

The following relations are evident:

$$\mathcal{Z}_j = \prod_{m=2^{j-1}+1}^{2^j} E\, e^{it\frac{X_m}{6_n}} = \prod_{m=1}^{2^{j-1}} E\, e^{it\frac{X_m}{6_n}} = \prod_{m=0}^{j-1} \mathcal{Z}_m ,$$

$$E\, e^{it Y_j} = E\, e^{it\sum_{k=2^{j-1}+1}^{2^j}\frac{X_k}{6_n}} = E\, e^{it\sum_{k=1}^{2^{j-1}}\frac{X_k}{6_n}} =$$

$$= E\, e^{it\hat{S}_{j-1}} = E\prod_{k=0}^{j-1} e^{it Y_k} , \quad j = \overline{2, \tau} . \tag{5.5.1}$$

Further we can write

$$\mathcal{U}_j \leq \left| E\prod_{s=0}^{j} e^{it Y_s} - \prod_{s=0}^{j} E\, e^{it Y_s} \right| + \left| \prod_{s=0}^{j} E\, e^{it Y_s} - \prod_{s=0}^{j} \mathcal{Z}_s \right| \leq$$

$$\leq \alpha_j + \sum_{s=0}^{j} \left| E\, e^{it Y_s} - \mathcal{Z}_s \right| = \alpha_j + \sum_{s=1}^{j} \left| E\, e^{it Y_s} - \mathcal{Z}_s \right| =$$

$$= \alpha_j + \sum_{s=0}^{j-1} \left| E\, e^{it Y_{s+1}} - \mathcal{Z}_{s+1} \right| =$$

$$= \alpha_j + \sum_{s=0}^{j-1} \left| E\prod_{k=0}^{s} e^{it Y_k} - \prod_{k=0}^{s} \mathcal{Z}_k \right| = \alpha_j + \sum_{s=0}^{j-1} \mathcal{U}_s , \quad j = \overline{2, \tau} .$$

Thus
$$\mathcal{U}_0 = 0, \quad \mathcal{U}_1 = \alpha_1 , \quad \mathcal{U}_j \leq \alpha_j + \sum_{s=0}^{j-1} \mathcal{U}_s , \quad j = \overline{2, \tau} . \tag{5.5.2}$$

Solving this recursive inequality we obtain
$$\mathcal{U}_0 = 0, \quad \mathcal{U}_1 = \alpha_1 , \quad \mathcal{U}_j \leq \alpha_j + \sum_{s=0}^{j-1} 2^{j-s-1} \alpha_s , \quad j = \overline{2, \tau} . \tag{5.5.3}$$

Now let us estimate α_j , $j = \overline{2, \tau}$. Using inequality (3.3.7) we have

$$\alpha_j \leq \sum_{m=0}^{j-1} \sum_{s=m+1}^{j} \left| E(e^{itY_m}-1)(e^{itY_s}-1) \prod_{k=s+1}^{j} e^{itY_k} - \right.$$

$$\left. - E(e^{itY_m}-1) E(e^{itY_s}-1) \prod_{k=s+1}^{j} e^{itY_k} \right| =$$

$$= \sum_{m=0}^{j-1} \left| E(e^{itY_m}-1)(e^{itY_{m+1}}-1) \prod_{k=m+2}^{j} e^{itY_k} - \right.$$

$$\left. - E(e^{itY_m}-1) E(e^{itY_{m+1}}-1) \prod_{k=m+2}^{j} e^{itY_k} \right| +$$

$$+ \sum_{m=0}^{j-1} \sum_{s=m+2}^{j} \left| E(e^{itY_m}-1)(e^{itY_s}-1) \prod_{k=s+1}^{j} e^{itY_k} - \right.$$

$$\left. - E(e^{itY_m}-1) E(e^{itY_s}-1) \prod_{k=s+1}^{j} e^{itY_k} \right|.$$

Taking into consideration the definition of ρ -mixing coefficient we can write

$$\sum_{m=0}^{j-1} \sum_{s=m+2}^{j} \left| E(e^{itY_m}-1)(e^{itY_s}-1) \prod_{k=s+1}^{j} e^{itY_k} - \right.$$

$$\left. - E(e^{itY_m}-1) E(e^{itY_s}-1) \prod_{k=s+1}^{j} e^{itY_k} \right| \leq$$

$$\leq \sum_{m=0}^{j-1} \sum_{s=m+2}^{j} E^{\frac{1}{2}} |e^{itY_m}-1|^2 E^{\frac{1}{2}} |e^{itY_s}-1|^2 \rho(2^{s-2}q) \leq$$

$$\leq t^2 \sum_{m=0}^{j-1} \sum_{s=m+2}^{j} E^{\frac{1}{2}} |Y_m|^2 E^{\frac{1}{2}} |Y_s|^2 \rho(2^{s-2}q) \leq$$

$$\leq \frac{C_1 t^2 p}{\sigma_n^2} \sum_{m=0}^{j-1} 2^{\frac{m}{2}} \sum_{s=m+2}^{j} 2^{\frac{s}{2}} \rho(2^{s-2}q) \leq$$

$$\leq \frac{C_1 t^2 p}{\sigma_n^2} \sum_{s=0}^{j} 2^{\frac{s}{2}} \rho(2^{s-2}q) \sum_{m=1}^{s} 2^{\frac{m}{2}} \leq$$

$$\leq \frac{C_2 t^2 p}{\sigma_n^2} \sum_{s=0}^{j} 2^s \rho(2^{s-1} q), \qquad 0 < C_1, C_2 < \infty.$$

Let's denote

$$Y_j' = \frac{1}{\sigma_n} \sum_{m=2^{j-1}+1}^{2^{j}-2^{[\frac{j}{2}]}} X_m, \qquad Y_j'' = \frac{1}{\sigma_n} \sum_{m=2^{j-1}+2^{[\frac{j}{2}]}+1}^{2^{j}} X_m,$$

$$j = \overline{1, \tau}.$$

Hence we have

$$\sum_{m=0}^{j-1} \left| E(e^{itY_m}-1)(e^{itY_{m+1}}-1) \prod_{s=m+2}^{j} e^{itY_s} - \right.$$

$$\left. - E(e^{itY_m}-1) E(e^{itY_{m+1}}-1) \prod_{s=m+2}^{j} e^{itY_s} \right| \leq$$

$$\leq \sum_{m=0}^{j-1} \left| E(e^{itY_m}-e^{itY_m'})(e^{itY_{m+1}}-e^{itY_{m+1}''}) \prod_{s=m+2}^{j} e^{itY_s} - \right.$$

$$\left. - E(e^{itY_m}-e^{itY_m'}) E(e^{itY_{m+1}}-e^{itY_{m+1}''}) \prod_{s=m+2}^{j} e^{itY_s} \right| +$$

$$+ \sum_{m=0}^{j-1} \left| E(e^{itY_m'}-1)(e^{itY_{m+1}}-1) \prod_{s=m+2}^{j} e^{itY_s} - \right.$$

$$\left. - E(e^{itY_m'}-1) E(e^{itY_{m+1}}-1) \prod_{s=m+2}^{j} e^{itY_s} \right| +$$

$$+ \sum_{m=0}^{j-1} \left| E(e^{itY_m}-e^{itY_m'})(e^{itY_{m+1}''}-1) \prod_{s=m+2}^{j} e^{itY_s} - \right.$$

$$\left. - E(e^{itY_m}-e^{itY_m'}) E(e^{itY_{m+1}''}-1) \prod_{s=m+2}^{j} e^{itY_s} \right| \leq$$

$$\leq \sum_{m=0}^{j-1} E^{\frac{1}{2}} \left| e^{itY_m}-e^{itY_m'} \right|^2 E^{\frac{1}{2}} \left| e^{itY_m}-e^{itY_{m+1}''} \right|^2 \rho(q) +$$

$$+ \sum_{m=0}^{j-1} E^{\frac{1}{2}} |e^{itY_m'} - 1|^2 \, E^{\frac{1}{2}} |e^{itY_{m+1}} - 1|^2 \, \rho \left(2^{\left[\frac{m}{2}\right]} q\right) +$$

$$+ \sum_{m=0}^{j-1} E^{\frac{1}{2}} |e^{itY_m} - e^{itY_m'}|^2 \, E^{\frac{1}{2}} |e^{itY_{m+1}''} - 1|^2 \, \rho \left(2^{\left[\frac{m}{2}\right]} q\right) \leq$$

$$\leq \frac{C_1' t^2 P}{\sigma_n^2} \left(\sum_{m=0}^{j} 2^{\frac{m}{4}} 2^{\frac{m}{4}} \rho(q) + \sum_{m=0}^{j} 2^{\frac{m}{2}} 2^{\frac{m}{2}} \rho \left(2^{\left[\frac{m}{2}\right]} q\right) + \right.$$

$$+ \left. \sum_{m=0}^{j} 2^{\frac{m}{4}} 2^{\frac{m}{2}} \rho \left(2^{\left[\frac{m}{2}\right]} q\right) \right) \leq$$

$$\leq \frac{C_2' t^2 P}{\sigma_n^2} \left(2^{\frac{j}{2}} \rho(q) + \sum_{m=0}^{j} 2^{m} \rho \left(2^{\left[\frac{m}{2}\right]} q\right) \right),$$

$$0 < C_1', \ C_2' < \infty .$$

Using this inequality and (5.5.3) we obtain

$$U_j \leq \frac{C_2 t^2 P}{\sigma_n^2} \left[\sum_{m=0}^{j} 2^{m} \rho \left(2^{\left[\frac{m}{2}\right]} q\right) + 2^{\frac{j}{2}} \rho(q) + \right.$$

$$+ 2^j \sum_{s=0}^{j-1} \frac{1}{2^{s+1}} \sum_{m=0}^{s} 2^{m} \rho \left(2^{\left[\frac{m}{2}\right]} q\right) + 2^j \sum_{s=0}^{j-1} \frac{2^{\frac{s}{2}}}{2^{s+1}} \rho(q) \left. \right] =$$

$$= \frac{C_2 t^2 P}{\sigma_n^2} \left[\sum_{m=0}^{j} 2^{m} \rho \left(2^{\left[\frac{m}{2}\right]} q\right) + 2^{\frac{j}{2}} \rho(q) + \right.$$

$$+ 2^j \sum_{m=0}^{j} 2^{m} \rho \left(2^{\left[\frac{m}{2}\right]} q\right) \sum_{s=m+1}^{j} \frac{1}{2^{s}} + 2^j \rho(q) \left. \right] .$$

Finally we have

$$\mathcal{U}_j \le \frac{C_3 t^2 2^j P}{\sigma_n^2} \left(\sum_{k=0}^{\infty} \rho(2^{[\frac{k}{2}]} q) + \rho(q) \right), \quad j = \overline{1, \tau}. \qquad (5.5.4)$$

Similarly we can estimate the following quantities

$$V_j = D\left(\sum_{k=0}^{j} \overline{Y}_k \right), \quad j = \overline{0, \tau}.$$

We have

$$V_j \le \beta_j + \sum_{k=0}^{j-1} V_k , \quad j = \overline{1, \tau}, \qquad (5.5.5)$$

where

$$\beta_j = 2 \sum_{k=1}^{j-1} \sum_{t=k+1}^{j} |E \overline{Y}_k \overline{Y}_t| .$$

Hence

$$V_j \le \beta_j + \sum_{k=1}^{j-1} 2^{j-k-1} \beta_k . \qquad (5.5.6)$$

Let's estimate β_j, $j = \overline{1, \tau}$. We can write

$$\beta_j \le \sum_{k=1}^{j-1} |E \overline{Y}_k \overline{Y}_{k+1}| + \sum_{k=1}^{j-1} \sum_{t=k+2}^{j} |E \overline{Y}_k \overline{Y}_t|. \qquad (5.5.7)$$

Further

$$\sum_{k=1}^{j-1} \sum_{t=k+2}^{j} |E \overline{Y}_k \overline{Y}_t| \le$$

$$\le \sum_{k=1}^{j-1} \sum_{t=k+2}^{j} E^{\frac{1}{2}} |\overline{Y}_k|^2 E^{\frac{1}{2}} |\overline{Y}_t|^2 \rho(2^{[\frac{t}{2}]} q) \le$$

$$\le \frac{C_1 q}{\sigma_n^2} \sum_{k=1}^{j-1} \sum_{t=k+2}^{j} 2^{\frac{k}{2}} 2^{\frac{t}{2}} \rho(2^{[\frac{t}{2}]} q) \le$$

$$\le \frac{C_2 q}{\sigma_n^2} \sum_{t=1}^{j} 2^t \rho(2^{[\frac{t}{2}]} q).$$

Let's denote

$$\overline{Y}_j' = \frac{1}{\sigma_n} \sum_{m=2^{j-1}+1}^{2^j - 2^{[\frac{j}{2}]}} \overline{X}_m , \quad \overline{Y}_j'' = \frac{1}{\sigma_n} \sum_{m=2^{j-1}+2^{[\frac{j}{2}]}+1}^{2^j} \overline{X}_m , \quad j = \overline{2, \tau}.$$

118

Hence

$$\sum_{k=1}^{j-1} |E\,\overline{Y}_k\,\overline{Y}_{k+1}| \le \sum_{k=1}^{j-1} |E(\overline{Y}_k - \overline{Y}_k')(\overline{Y}_{k+1} - \overline{Y}_{k+1}'')| +$$

$$+\sum_{k=1}^{j-1} |E(\overline{Y}_k - \overline{Y}_k')\,\overline{Y}_{k+1}''| + \sum_{k=1}^{j-1} |E\,\overline{Y}_k'\,\overline{Y}_{k+1}| \le$$

$$\le \frac{C_3\,q}{\sigma_n^2}\left(2^{\frac{j}{2}}\rho(q) + \sum_{k=1}^{j} 2^k \rho\left(2^{[\frac{k}{2}]}q\right)\right), \quad j = \overline{1,\tau}.$$

Finally

$$\beta_j \le \frac{C_3\,q}{\sigma_n^2}\left(2^{\frac{j}{2}}\rho(q) + \sum_{k=1}^{j} 2^k \rho\left(2^{[\frac{k}{2}]}q\right)\right), \quad j = \overline{1,\tau}. \tag{5.5.8}$$

Substituting (5.5.8) into (5.5.6) we obtain

$$V_j \le \frac{C_4\,q\,2^j}{\sigma_n^2}\left(\rho(q) + \sum_{k=1}^{\infty} \rho\left(2^{[\frac{k}{2}]}q\right)\right). \tag{5.5.9}$$

Now we show that for any standard pair of functions p, q the following relation is valid

$$\left| E\prod_{j=0}^{k} e^{it\frac{X_j + \overline{X}_j}{\sigma_n}} - \prod_{j=0}^{k} E\,e^{it\frac{X_j}{\sigma_n}} \right| \to 0, \quad n \to \infty.$$

We have

$$\left| E\prod_{j=0}^{k} e^{it\frac{X_j + \overline{X}_j}{\sigma_n}} - \prod_{j=0}^{k} E\,e^{it\frac{X_j}{\sigma_n}} \right| \le \left| E\prod_{j=0}^{k} e^{it\frac{X_j + \overline{X}_j}{\sigma_n}} - E\prod_{j=0}^{k} e^{it\frac{X_j}{\sigma_n}} \right| +$$

$$+\left| E\prod_{j=0}^{k} e^{it\frac{X_j}{\sigma_n}} - \prod_{j=0}^{k} E\,e^{it\frac{X_j}{\sigma_n}} \right|. \tag{5.5.10}$$

Let us estimate the first term on the right-hand side of (5.5.10). We can write for $k = 2^\tau$

$$\left| E\prod_{j=0}^{k} e^{it\frac{X_j + \overline{X}_j}{\sigma_n}} - E\prod_{j=0}^{k} e^{it\frac{X_j}{\sigma_n}} \right| \le$$

$$\le \left| E\prod_{j=0}^{\tau} e^{it(Y_j + \overline{Y}_j)} - E\prod_{j=0}^{\tau} e^{it\,Y_j} \right| \le$$

$$\leq \sum_{j=0}^{n} \left| E\, e^{it(Y_j + \overline{Y}_j)} - E\, e^{it Y_j} \right| \leq$$

$$\leq |t| \sum_{j=0}^{n} E\, |\overline{Y}_j| \leq |t| \sum_{j=0}^{n} \sqrt{V_j} \; .$$

Using (5.5.9) we obtain

$$\left| E\prod_{j=0}^{K} e^{it\frac{X_j + \overline{X}_j}{\sigma_n}} - E\prod_{j=0}^{K} e^{it\frac{X_j}{\sigma_n}} \right| \leq \frac{C_1 |t| \sqrt{q}}{\sigma_n} \sum_{j=1}^{n} 2^{\frac{j}{2}} (\rho(q) +$$

$$+ \sum_{K=1}^{\infty} \rho(2^{[\frac{K}{2}]} q))^{\frac{1}{2}} \leq C_2 |t| \sqrt{\tfrac{q}{p}} \, (\rho(q) + \sum_{K=1}^{\infty} \rho(2^{[\frac{K}{2}]} q))^{\frac{1}{2}} .$$

Further

$$\left| E\prod_{j=1}^{K} e^{it\frac{X_j}{\sigma_n}} - \prod_{j=1}^{K} E\, e^{it\frac{X_j}{\sigma_n}} \right| \leq \left| E\prod_{j=1}^{n} e^{it Y_j} - \prod_{j=1}^{n} E\, e^{it Y_j} \right| +$$

$$+ \left| \prod_{j=1}^{n} E\, e^{it Y_j} - \prod_{j=1}^{n} z_j \right| \leq \sum_{j=1}^{n} \alpha_j + \sum_{j=1}^{n} u_j \leq$$

$$\leq \frac{C_4 t^2 p}{\sigma_n^2} \sum_{j=0}^{n} 2^{j} \left(\sum_{K=0}^{\infty} \rho(2^{[\frac{K}{2}]} q) + \rho(q) \right).$$

Finally, since by Theorem 3.1.6 $\sigma_n^2 \sim \sigma^2 n$ as $n \to \infty$, $\sigma^2 > 0$ then we find

$$\left| E\prod_{j=1}^{K} e^{it\frac{X_j + \overline{X}_j}{\sigma_n}} - \prod_{j=1}^{K} E\, e^{it\frac{X_j}{\sigma_n}} \right| \leq$$

$$\leq C_5 \Big[|t| \sqrt{\tfrac{q}{p}} \, \big(\sum_{K=0}^{\infty} \rho(2^{[\frac{K}{2}]} q) + \rho(q) \big)^{\frac{1}{2}} + t^2 \big(\sum_{K=0}^{\infty} \rho(2^{[\frac{K}{2}]} q) + \rho(q) \big) \Big] .$$

Now consider the case of general K, $K = \sum_{j=0}^{n} v_j 2^{n-j}$, $v_0 \neq 0$. Denote

$$W_j = \sum_{t=1}^{2^j} \frac{X_t}{\sigma_n} , \qquad \overline{W}_j = \sum_{t=1}^{2^j} \frac{\overline{X}_t}{\sigma_n} \; .$$

Then

$$\left| E\prod_{j=0}^{K} e^{it\frac{X_j+\overline{X}_j}{6_n}} - E\prod_{j=0}^{K} e^{it\frac{X_j}{6_n}}\right| \le \sum_{j=0}^{\imath} \nu_j \left| E e^{it(W_j+\overline{W}_j)} - E e^{it W_j}\right|$$

$$\le |t| \sum_{j=0}^{\imath} \nu_j \, E|\overline{W}_j| \le |t| \sum_{j=0}^{\imath} \nu_j \sqrt{Y_j} \le$$

$$\le \frac{\overline{C}_1 |t| \sqrt{q}}{6_n} \sum_{j=0}^{\imath} \nu_j \, 2^{\frac{j}{2}} \Big(\rho(q) + \sum_{K=1}^{\infty} \rho(2^{\lceil\frac{K}{2}\rceil} q)\Big)^{\frac{1}{2}} \le$$

$$\le \overline{C}_2 \, |t| \, \sqrt{\frac{q}{p}} \, \Big(\rho(q) + \sum_{K=1}^{\infty} \rho(2^{\lceil\frac{K}{2}\rceil} q)\Big)^{\frac{1}{2}}.$$

Further

$$\left| E\prod_{j=0}^{K} e^{it\frac{X_j}{6_n}} - \prod_{j=0}^{K} E e^{it\frac{X_j}{6_n}}\right| \le \sum_{j=0}^{\imath} \nu_j \left| E\prod_{s=0}^{2^j} e^{it\frac{X_s}{6_n}} - \prod_{s=0}^{2^j} E e^{it\frac{X_s}{6_n}}\right| \le$$

$$\le \frac{C_4 t^2 p}{6_n^2} \sum_{j=0}^{\imath} \nu_j \, 2^j \Big(\sum_{K=0}^{\infty} \rho(2^{\lceil\frac{K}{2}\rceil} q) + \rho(q)\Big).$$

Hence

$$\left| E\prod_{j=0}^{K} e^{it\frac{X_j+\overline{X}_j}{6_n}} - \prod_{j=0}^{K} E e^{it\frac{\overline{X}_j}{6_n}}\right| \le$$

$$\le C_5 \Big[|t| \sqrt{\frac{q}{p}} \Big(\sum_{K=0}^{\infty} \rho(2^{\lceil\frac{K}{2}\rceil} q) + \rho(q)\Big)^{\frac{1}{2}} +$$

$$+ t^2 \Big(\sum_{K=0}^{\infty} \rho(2^{\lceil\frac{K}{2}\rceil} q) + \rho(q)\Big)\Big] \to 0, \quad n \to \infty. \tag{5.5.11}$$

Theorem 5.5.1 is proven.

The conditions of this theorem are as weak as they are permitted to be. Bradley (/10/,/12/) constructed the ρ-mixing, stationary Gaussian sequences having only second order moments and such that

$$n^{-1}\sigma_n^2 \to 0, \quad \sigma_n^2 \to \infty, \quad n \to \infty$$

and

$$\rho(n) = O(\log n)^{-1}, \quad n \to \infty .$$

The theorem given below directly follows from Corollary 5.3.2 and Lemma 3.2.2.

<u>Theorem 5.5.2</u>(Ibragimov /65/).Suppose a centered s.r.p. $\xi_t,\ t \in \mathbb{Z}^1$ satisfies ρ -mixing condition, $E|\xi_0|^{2+\delta} < \infty$ for some $\delta > 0$ and

$$\sigma_n^2 \to \infty, \quad n \to \infty.$$

Then for the process the functional form of the c.l.th. holds.

Now we formulate some conditions under which the functional form of the c.l.th. holds.

<u>Theorem 5.5.3</u>.Let $\xi_t,\ t \in \mathbb{Z}^1$ be a ρ -mixing sequence of random variables satisfying one of the following three groups of conditions:

A)   1) $\left(\xi_t^2,\ t \in \mathbb{Z}^1 \right)$ is uniformly integrable ,

          2) $\sum\limits_{n=1}^{\infty} \rho(2^n) < \infty$,

          3) $\lim\limits_{n \to \infty} \varphi(n) < 1$,

B)   1) $\left(\xi_t^2,\ t \in \mathbb{Z}^1 \right)$ is uniformly integrable ,

          2) $\sum\limits_{n=1}^{\infty} \rho^{\frac{1}{2}}(2^n)$,

C)   1) $DS_n = n\, h(n)$,

      where $h(n)$ is a slowly varying function,

          2) $\sup\limits_{t \in \mathbb{Z}^1} E|\xi_t|^{2+\delta}$ for some $\delta > 0$,

Then for the process the functional form of the c.l.th. holds.

Note, that Part A and Part B of Theorem 5.5.3 were proved in /65/
Part C is essentially contained in /65/.

__Theorem 5.5.4.__ Suppose a centered s.r.p. ξ_t, $t \in \mathbb{Z}^1$ satisfies
ρ -mixing condition and

$$\rho(n) = O\left(\frac{1}{\ln^{4+\omega} n}\right), \quad \omega > 0, \quad E|\xi_0|^{2+\delta} < \infty, \quad \delta > 0.$$

If $\sigma_n^2 \to \infty$ as $n \to \infty$, then

$$\Delta_n = \sup_{x \in R^1} \left| P\left(\frac{S_n}{\sigma_n} < x\right) - \Phi(x)\right| = O\left(\frac{1}{\ln^{1+\rho} n}\right), \quad \rho > 0.$$

__Proof.__ Using Theorems 1.6.1 , 1.6.4 and the inequality (5.5.11) we
have

$$\Delta_n \leq C\left(\frac{1}{T} + T\sqrt{\frac{q}{p}}\;\sqrt{\rho(q) + \sum_{K=1}^{\infty} \rho(2^{[\frac{K}{2}]}q)}\;+\right.$$

$$\left.+ T^2\left(\rho(q) + \sum_{K=1}^{\infty} \rho(2^{[\frac{K}{2}]}q)\right) + \varepsilon + \frac{n}{p\sigma_n^2}\int\limits_{|S_p| \geq \varepsilon\sigma_n} S_p^2\, P(d\omega)\right). \tag{5.5.12}$$

According to Theorem 3.1.6 and Lemma 3.2.2 we can write

$$\frac{n}{p\sigma_n^2}\int\limits_{|S_p| \geq \varepsilon\sigma_n} S_p^2\, P(d\omega) = O\left(\left(\frac{p}{\varepsilon_n}\right)^\delta\right), \quad \delta > 0.$$

Next we have

$$\sum_{K=1}^{\infty} \rho(2^{[\frac{K}{2}]}q) \leq C_1 \sum_{K=1}^{\infty} \frac{1}{(K+\ln q)^{4+\omega}} \leq C_1\left(\sum_{K \leq \ln q} \frac{1}{(\ln q)^{4+\varepsilon}}\;+\right.$$

$$+ \sum_{K \geq \ln q} \frac{1}{K^{4+\varepsilon}} \leq \frac{C_2}{(\ln q)^{3+\omega}}, \quad \omega > 0.$$

Now putting in (5.5.12)

$$T = \ell n^{1+\rho'} n, \quad q(n) = n^{\alpha}, \quad p(n) = n^{\beta}, \quad 0 < \alpha < \beta < 1,$$

$$\varepsilon = 1/\ell n^{1+\rho'} n, \qquad \omega > 2\rho'$$

we obtain

$$\Delta_n = \frac{C}{\ell n^{1+\rho} n}, \qquad \rho = \omega - 2\rho'.$$

__Theorem 5.5.5.__ Let ξ_t, $t \in \mathbb{Z}^1$ be a centered ρ-mixing s.r.p. such that

$$\rho(n) = O\left(\frac{1}{\ell n^{4+\varepsilon} n}\right), \quad \omega > 0, \quad E|\xi_0|^{2+\delta} < \infty, \quad \delta > 0.$$

If $\;\sigma_n^2 \to \infty\;$ as $\;n \to \infty\;$ then the process satisfies the law of iterated logarithm.

__Proof.__ Obviously , it is sufficient to verify the condition 2) of Theorem 5.4.1. Using Theorem 3.3.4 we can write

$$P\left(\max_{1 \le j \le n} |S_j| \ge \theta \chi(n)\right) + \frac{\rho(P)}{1-\eta} \sum_{i=0}^{K-2} P\left(|S_n - S_{(i+2)P}| \ge \theta \chi(n)\right) +$$

$$+ \frac{1}{1-\eta} \cdot \frac{n}{P} \cdot P\left(\sum_{i=1}^{P} |\xi_i| \ge \theta \chi(n)\right), \qquad K = \left[\frac{n}{P}\right]$$

for sufficiently large n .

Putting

$$P = \left[n^{\delta/2(1+\delta)} \ell n^{-1} n\right]$$

we have

$$\frac{n}{P} P\left(\sum_{j=1}^{2P} |\xi_j| \ge \tfrac{1}{2} \theta \chi(n)\right) = O\left(\frac{1}{\ell n^{1+\rho} n}\right), \qquad \rho > 0$$

(see the proof of Theorem 5.4.3) and

124

$$P\left(|S_n| \geqslant \frac{\beta\chi(n)}{2}\right) = O\left(\frac{1}{\ln^{1+\rho}n}\right), \qquad \rho > 0$$

(see the proof of Theorem 5.4.2).
Further

$$\sum_{i=0}^{K-2} P\left(|S_n - S_{(i+2)p}| \geqslant \beta\chi(n)\right) \leqslant \sum_{i=1}^{K} P\left(|S_i| \geqslant \beta\chi(n)\right) \leqslant$$

$$\leqslant C \sum_{i=1}^{K} \frac{i^{\frac{2+\delta}{2}}}{n^{\frac{2+\delta}{2}}} = C_1 \frac{K^{\frac{2+\delta}{2}}}{n^{\frac{2+\delta}{2}}} = C_1 \frac{n^2 \cdot n^{\frac{\delta}{2}}}{n^{1+\frac{\delta}{2}}p^{2+\frac{\delta}{2}}} =$$

$$= C_1 \frac{n}{p^{\frac{4+\delta}{2}}} = O\left(\frac{1}{\ln^{1+\rho}n}\right).$$

The theorem is proven.

5.6 Limit Theorems for Stationary Random Processes Satisfying the φ -mixing Condition

In /66/ Ibragimov suggested the following conjecture: for a s.r.p. ξ_t, $t \in \mathbb{Z}^1$ satisfies φ -mixing condition and such that $E\xi_0^2 < \infty$, $\sigma_n^2 \to \infty$ as $n \to \infty$ the c.l.th. is valid. The theorem given below is an essential step to establish towards Ibragimov's conjecture.

Theorem 5.6.1(/100/).Let ξ_t, $t \in \mathbb{Z}^1$ be a centered s.r.p. such that

$$E\xi_0^2 < \infty, \quad \sigma_n^2 \to \infty \quad \text{as} \quad n \to \infty \quad \text{and}$$

$$\lim_{n \to \infty} \frac{n}{\sigma_n^2} \int_{|\xi_0| \geqslant \varepsilon\sigma_n} \xi_0^2 \, P(d\omega) = 0, \quad \varepsilon > 0.$$

Then for the process the c.l.th. holds.

Proof. According to Theorem 3.1.3 we have $\sigma_n^2 = n\,h(n)$, where $h(n)$ is a slowly varying function. Thus take into account Theorem 5.1.1 we conclude that for the proving Theorem 5.6.1 it is sufficient to show that

$$\frac{n}{P\sigma_n^2} \int\limits_{|S_P| \geq \varepsilon\sigma_n} S_P^2 \, P(d\omega) \to 0 \,, \qquad\qquad n \to \infty \,.$$

We have

$$\frac{1}{P\,h(n)} \int\limits_{|S_P| \geq \varepsilon\sigma_n} |S_P|^2 P(d\omega) = \frac{1}{P\,h(n)} \left[\varepsilon^2 \sigma_n^2 \, P\big(|S_P| \geq \varepsilon\sigma_n\big) + \right.$$

$$\left. + \int\limits_{x \geq \varepsilon^2 \sigma_n^2} P\big(|S_P| \geq \sqrt{x}\,\big) dx \right] . \tag{5.6.1}$$

Let's estimate the first term in (5.6.1). Using Theorem 3.3.3 we can write for sufficiently large $q \in \mathbb{N}$

$$\frac{\varepsilon^2 n}{P} \, P\big(|S_P| \geq \varepsilon\sigma_n\big) \leq$$

$$\leq \frac{\varepsilon^2 n}{P} \left[\frac{\eta}{1-\eta} P\big(|S_P| \geq \frac{\varepsilon\sigma_n}{3}\big) + \frac{1}{1-\eta} P\big(\max_{1 \leq j \leq P} |\xi_j| \geq \frac{\varepsilon\sigma_n}{3q}\big) \right] ,$$

$$\eta = \varphi(q) + \max_{1 \leq i \leq P} P\big(|S_P - S_i| \geq \frac{\varepsilon\sigma_n}{3}\big) < \frac{1}{2} .$$

Further

$$\varepsilon^2 \frac{n}{P} \, P\big(|S_P| \geq \frac{\varepsilon\sigma_n}{3}\big) \leq$$

$$\leq \frac{\varepsilon^2}{1-\eta} \left[\frac{n\,\varphi(q)}{P} P\big(|S_P| \geq \frac{\varepsilon\sigma_n}{3}\big) + \frac{n}{P} P\big(\max_{1 \leq j \leq P} |\xi_j| \geq \frac{\varepsilon\sigma_n}{3}\big) + \right.$$

$$\left. + \frac{n}{P} \max_{1 \leq i \leq P} P\big(|S_P - S_i| \geq \frac{\varepsilon\sigma_n}{3}\big) P\big(|S_P| \geq \frac{\varepsilon\sigma_n}{3}\big) \right. .$$

We have

$$\frac{n}{P} \varphi(q) P\big(|S_P| \geq \frac{\varepsilon\sigma_n}{3}\big) \leq \frac{C_1}{\varepsilon^2} \frac{n}{P} \frac{DS_P}{DS_n} \varphi(q) \leq C_2 \frac{h(P)}{h(n)} \varphi(q) .$$

Obviously there exists a standard pair of functions P, q such that

$$\frac{h(p)}{h(n)} \to 1, \qquad n \to \infty$$

and for which all requirements of Theorem 5.5.1 are valid.
Therefore under this choice of $\quad p = p(n) \quad$ we obtain

$$\frac{n}{p} \, \varphi(q) \, P\left(|S_p| \geqslant \frac{\varepsilon \sigma_n}{3}\right) \to 0, \qquad n \to \infty.$$

Further

$$\frac{n}{p} \max_{1 \leqslant i \leqslant p} P\left(|S_p - S_i| \geqslant \frac{\varepsilon \sigma_n}{3}\right) P\left(|S_p| \geqslant \frac{\varepsilon \sigma_n}{3}\right) \leqslant$$

$$\leqslant C_1 \frac{n}{p} \max_{1 \leqslant i \leqslant p} \frac{DS_i}{\sigma_n^2} \cdot \frac{DS_p}{\sigma_n^2} = C_1 \frac{n}{p} \frac{Ph(p)}{n h(n)} \frac{\max\limits_{1 \leqslant i \leqslant p} i \, \frac{h(i)}{h(n)}}{n} \leqslant$$

$$\leqslant C_1 \frac{h(p)}{h(n)} \cdot \frac{\max\limits_{1 \leqslant i \leqslant p} i \left(\frac{n}{i}\right)^{\tau_1}}{n} \leqslant C_1 \left(\frac{n}{p}\right)^{\tau_2} \frac{\max\limits_{1 \leqslant i \leqslant p} i^{1-\tau_1}}{n^{1-\tau_1}} \leqslant$$

$$\leqslant C_1 \left(\frac{n}{p}\right)^{\tau_2} \left(\frac{p}{n}\right)^{1-\tau_1}.$$

If $\quad 1 - \tau_1 > \tau_2 \quad$ then the latter tends to zero as $n \to \infty$
We also can write

$$\frac{n}{p} P\left(\max_{1 \leqslant j \leqslant p} |\xi_j| \geqslant \frac{\varepsilon \sigma_n}{3q}\right) \leqslant \frac{n}{p} \sum_{j=1}^{p} P\left(|\xi_j| \geqslant \frac{\varepsilon \sigma_n}{3q}\right) \leqslant$$

$$\leqslant \frac{n}{p} \frac{p}{n h(n)} \int\limits_{|\xi_0| \geqslant \varepsilon \sigma_n q^{-1}} |\xi_0|^2 P(dw) \to 0, \qquad n \to \infty.$$

Further using Theorem 3.3.1 we obtain

$$\frac{1}{p h(n)} \int\limits_{x \geqslant \varepsilon^2 \sigma_n^2} P\left(|S_p| \geqslant \sqrt{x}\right) dx \leqslant \frac{1}{p h(n)(1-\eta)} \Bigg[\int\limits_{x \geqslant \varepsilon^2 \sigma_n^2} \eta P\left(|S_p| \geqslant \frac{\sqrt{x}}{3}\right) dx +$$

$$+ \int\limits_{x \geqslant \varepsilon^2 \sigma_n^2} \eta P\left(\max_{1 \leqslant j \leqslant p} |\xi_j| \geqslant \frac{\sqrt{x}}{3q}\right) dx \Bigg] \leqslant$$

$$\leq 2 \left[\frac{\varphi(q)}{P\,h(n)} \int\limits_{x \geq \varepsilon^2 \sigma_n^2} P\left(|S_P| \geq \frac{\sqrt{x}}{3}\right) dx + \right.$$

$$+ \frac{1}{p\,h(n)} \int\limits_{x \geq \varepsilon^2 \sigma_n^2} \max_{1 \leq i \leq P} P\left(|S_P - S_i| \geq \frac{\sqrt{x}}{3}\right) P\left(|S_P| \geq \frac{\sqrt{x}}{3}\right) dx +$$

$$+ \left. \frac{1}{p\,h(n)} \int\limits_{x \geq \varepsilon^2 \sigma_n^2} P\left(\max_{1 \leq j \leq P} |\xi_j| \geq \frac{\sqrt{x}}{3q}\right) dx \right] \leq$$

$$\leq 2 \left[\frac{\varphi(q)Ph(P)}{P\,h(n)} + \frac{C_1}{p\,h(n)} \max_{1 \leq i \leq P} P\left(|S_P - S_i| \geq \varepsilon\sigma_n\right) \int\limits_{x \geq \varepsilon^2 \sigma_n^2} P\left(|S_P| \geq \frac{\sqrt{x}}{3}\right) dx + \right.$$

$$+ \left. \frac{1}{Ph(n)} \sum_{j=1}^{P} \int\limits_{|x| \geq \varepsilon^2 \sigma_n^2} P\left(|\xi_j| > \frac{\sqrt{x}}{3q}\right) dx \right].$$

Now we can write

$$\frac{1}{Ph(n)} \max_{1 \leq i \leq P} P\left(|S_P - S_i| \geq \varepsilon\sigma_n\right) \int\limits_{x \geq \varepsilon^2 \sigma_n^2} P\left(|S_P| \geq \frac{\sqrt{x}}{3}\right) dx \leq$$

$$\leq C_2 \frac{h(P)}{h(n)} \cdot \frac{\max\limits_{1 \leq i \leq P} i\,h(i)}{n\,h(n)} \to 0, \qquad n \to \infty.$$

Finally by the conditions of Theorem 5.6.1

$$\frac{1}{h(n)} \int\limits_{|x| \geq \varepsilon^2 \sigma_n^2} P\left(|\xi_j| > \frac{\sqrt{x}}{3q}\right) dx \leq \frac{C_3}{h(n)} \int\limits_{|\xi_0| \geq \varepsilon\sigma_n^2} \xi_0^2 P(d\omega) \to 0, \quad n \to \infty.$$

Theorem 5.6.1 is proven.

Theorem 5.6.2. Under conditions of Theorem 5.6.1 the functional form of the c.l.th. holds.

Proof. It is sufficient to verify the conditions of Corollary 5.3.1. Since $\sigma_n^{-1} S_n$, $n \in \mathbb{N}$, is asymptotically normal then $(\sigma_n^{-1} S_n)^2$, $n \in \mathbb{N}$, is uniformly integrable. Let's check the validity

of relation (5.3.2).
Using Theorem 3.3.2 we find

$$P\left(\max_{1\leq i\leq n}|S_i|\geq \lambda\sigma_n\right)\leq$$

$$\leq \frac{1}{1-\eta}P\left(|S_n|>\frac{\lambda\sigma_n}{3}\right)+\frac{1}{1-\eta}P\left(\max_{1\leq i\leq n}|\xi_i|\geq \frac{\lambda\sigma_n}{3p}\right). \tag{5.6.1}'$$

By virtue of the uniform integrability of $\left(\sigma_n^{-1}S_n\right)^2$, $n\in\mathbb{N}$ we have

$$\frac{1}{1-\eta}P\left(|S_n|>\frac{\lambda\sigma_n}{3}\right)\leq \frac{\varepsilon}{\lambda^2}\ ,\qquad \varepsilon>0,\quad \lambda>1$$

Let us consider the second term in (5.6.1)' and fix p_0 such that $\eta<\frac{1}{2}$. Hence

$$P\left(\max_{1\leq i\leq n}|\xi_i|\geq \frac{\lambda\sigma_n}{3p}\right)\leq \sum_{j=1}^{n}P\left(|\xi_j|>\frac{\lambda\sigma_n}{3p_0}\right)\leq$$

$$\leq \frac{np_0}{n\,h(n)}\int_{|\xi_0|>\frac{\lambda\sigma_n}{3p_0}}|\xi_0|\,P(d\omega)\to 0,\qquad n\to\infty.$$

Theorem 5.6.2 is proven.

<u>Theorem 5.6.3.</u> Suppose a centered s.r.p. ξ_t, $t\in\mathbb{Z}^1$ satisfies φ -mixing condition and

$$\varphi^{\frac{1}{2}}(n)=O\left(\frac{1}{\ell n^{4+\varepsilon}}\right),\quad \varepsilon>0,\qquad \int_{|\xi_0|>N}\xi_0^2\,P(d\omega)=O\left(\frac{1}{\ell n^{1+\omega}N}\right),\quad \omega>0$$

If $\sigma_n^2\to\infty$ $n\to\infty$, then there exists some constant $\rho>0$ such that
$$\Delta_n\leq O\left(\frac{1}{\ell n^{1+\rho}n}\right).$$

<u>Proof.</u> Using (5.5.12) we have

$$\Delta_n\leq c\left(\frac{1}{T}+T\sqrt{\frac{q}{p}}\ \sqrt{\varphi^{\frac{1}{2}}(q)+\sum_{k=1}^{\infty}\varphi^{\frac{1}{2}}\left(2^{\left[\frac{k}{2}\right]}q\right)}+\right.$$

$$+ T^2 \left(\varphi^{\frac{1}{2}}(q) + \sum_{K=1}^{\infty} \varphi^{\frac{1}{2}}(2^{[\frac{K}{2}]} q) \right) + \varepsilon_1 + \frac{n}{\rho \sigma_n^2} \int\limits_{|S_P| \geqslant \varepsilon_1 \sigma_n} S_P^2 \, P(d\omega) .$$

Looking through the proof of Theorem 5.6.1 we can write

$$\frac{n}{\rho \sigma_n^2} \int\limits_{|S_P| \geqslant \varepsilon_1 n} S_P^2 \, P(d\omega) \leqslant C \left(\varphi(q) + \frac{P}{n} + \int\limits_{|\xi_0| \geqslant \varepsilon_1 \sigma_n} \xi_0^2 \, P(d\omega) \right) .$$

Putting $\quad T = \ell n^{1+\rho}(n), \quad \varepsilon_1 = \frac{1}{\ell n^{1+\rho} n}, \quad P = n^{\alpha}, \quad q = n^{\beta},$
$$0 < \beta < \alpha < 1$$

we finish the proof.

Theorem 5.6.4. Suppose a centered s.r.p. ξ_t, $t \in \mathbb{Z}^1$ satisfies the φ -mixing condition and

$$\varphi^{\frac{1}{2}}(n) = O \left(\frac{1}{\ell n^{4+\varepsilon} n} \right), \quad \varepsilon > 0, \quad \int\limits_{|\xi_0| > N} \xi_0^2 \, P(d\omega) = O \left(\frac{1}{\ell n^{1+\omega} N} \right) .$$

If $\sigma_n^2 \to \infty$ as $n \to \infty$ then for the process the law of iterated logarithm holds.

Proof. It is sufficient to show that the condition 2) of Theorem 5.4.1 is fulfilled. Using Theorem 3.3.2 we obtain

$$P \left(\max_{1 \leqslant i \leqslant n} |S_i| \geqslant \theta \chi(n) \right) \leqslant \frac{1}{1-\eta} \left[P \left(|S_n| > \frac{\theta \chi(n)}{3} \right) + P \left(\max_{1 \leqslant i \leqslant n} |\xi_i| \geqslant \frac{\theta \chi(n)}{3\rho} \right) \right]$$

($\rho \in \mathbb{N}$ is such that $\eta < 1$).
By standard way

$$P \left(|S_n| > \frac{\theta \chi(n)}{3} \right) = O \left(\frac{1}{(\ell n \, n)^{1+\rho}} \right) .$$

Further

$$P \left(\max_{1 \leqslant i \leqslant n} |\xi_i| \geqslant \frac{\theta \chi(n)}{3\rho} \right) \leqslant \sum_{i=1}^{n} P \left(|\xi_i| \geqslant \frac{\theta \chi(n)}{3\rho} \right) \leqslant$$

$$\leqslant \frac{C n \rho^2}{n} \int\limits_{|\xi_0| \geqslant \frac{\theta \chi(n)}{3\rho}} |\xi_0|^2 \, P(d\omega) = O \left(\frac{1}{(\ell n \, n)^{1+\rho}} \right), \quad \rho < \omega .$$

The theorem is proven.

Note that in a paper of Peligrad /100/ there are also established limit theorems for the non-stationary case. Let's mention some of them

__Theorem 5.6.5__. Let ξ_t, $t \in \mathbb{Z}^1$, be a centered φ -mixing sequence of random variables having finite moments of second order. Suppose the following conditions are valid:

1) $\sigma_n^2 = n\,h(n)$, where $h(n)$ is a slowing varying function defined on R^1 ;

2) $\sup\limits_{m \geqslant 0,\, n \geqslant 1} \left\{ \dfrac{E(S_{m+n} - S_m)^2}{\sigma_n^2} \right\} < \infty$;

3) $\lim\limits_{n \to \infty} \dfrac{1}{\sigma_n^2} \sum\limits_{i=1}^{n} E\,\xi_i^2\, I\,(\xi_i^2 > \varepsilon \sigma_n^2) = 0$ for every $\varepsilon > 0$.

Then for the process ξ_t, $t \in \mathbb{Z}^1$, the functional form of the c.l.th. holds.

__Corollary 5.6.1__. Let ξ_t, $t \in \mathbb{Z}^1$, be a centered second-order stationary φ -mixing sequence with $E\,\xi_0^2 < \infty$, $\sigma_n^2 \to \infty$ and

$$\lim\limits_{n \to \infty} \dfrac{1}{\sigma_n^2} \sum\limits_{i=1}^{n} E\,\xi_i^2\, I\,(\xi_i^2 > \varepsilon \sigma_n^2) = 0 \qquad \text{for every } \varepsilon > 0.$$

Then for the process ξ_t, $t \in \mathbb{Z}^1$, the functional form of the c.l.th. holds. If in addition $\varphi(1) < 1$ the converse is true.

Theorem 5.6.2 is also a consequence of Theorem 5.6.5.

5.7 Limit Theorems for ψ -mixing Stationary Random Processes

__Theorem 5.7.1__. (/86/ , /99/). Let ξ_t, $t \in \mathbb{Z}^1$ be a centered s.r.p. such that $E\,\xi_0^2 < \infty$, $\psi(n) \to 0$ as $n \to \infty$. If

$$\sigma_n^2 \geqslant cn, \quad c > 0 .$$

Then for the process the c.l.th. holds.

Of course, this theorem is a consequence of Theorem 5.6.1, but here we give another proof of this result which may be usefull in limit theorems for the random fields.

__Proof of Theorem 5.7.1__. It is sufficient to show that for some function $p = p(n)$ such that $p(n) \to \infty$, $p(n) = 0(n)$ as $n \to \infty$ the following relation is valid:

$$\lim_{n \to \infty} \frac{n}{P(n)\sigma_n^2} \int_{|S_P| \geqslant \varepsilon \sigma_n} S_P^2 \, P(d\omega) = 0, \quad \varepsilon > 0. \tag{5.7.1}$$

Let's estimate the integral

$$E \, S_P^2 \, I\left(|S_P| \geqslant \varepsilon \sigma_n\right).$$

We have

$$E \, S_P^2 \, I\left(|S_P| \geqslant \varepsilon \sigma_n\right) = E\left(\sum_{j=1}^{P} \xi_j^2\right) I\left(|S_P| \geqslant \varepsilon \sigma_n\right) +$$
$$+ E\left(\sum_{\substack{t,s=1 \\ t \neq s}}^{P} \xi_t \xi_s\right) I\left(|S_P| \geqslant \varepsilon \sigma_n\right). \tag{5.7.2}$$

Further

$$\left| E\left(\sum_{\substack{t,s=1 \\ t \neq s}}^{P} \xi_t \xi_s\right) I\left(|S_P| \geqslant \varepsilon \sigma_n\right)\right| \leqslant \left(\varepsilon \sigma_n\right)^{-1} E \left|\sum_{\substack{t,s=1 \\ t \neq s}}^{P} \xi_t \xi_s\right| \left|\sum_{t=1}^{P} \xi_t\right|.$$

Using Lemma 3.2.3 we can write

$$\left| E\left(\sum_{\substack{t,s=1 \\ t \neq s}}^{P} \xi_t \xi_s\right) I\left(|S_P| \geqslant \varepsilon \sigma_n\right)\right| \leqslant C_1 \left(\varepsilon \sigma_n\right)^{-1} \sigma_P^3.$$

Since $\sigma_n^2 = n \, h(n)$, where $h(n)$ is a slowly varying function we have

$$\frac{n}{P\sigma_n^2} \left| E\left(\sum_{\substack{t,s=1 \\ t \neq s}}^{P} \xi_t \xi_s\right) I\left(|S_P| \geqslant \varepsilon \sigma_n\right)\right| \leqslant \frac{C_2 \, n \sigma_P^3}{P \sigma_n^3} \to 0, \quad n \to \infty.$$

Next

$$n P^{-1} \sigma_n^{-2} \left| E\left(\sum_{t=1}^{P} \xi_t^2\right) I\left(|S_P| \geqslant \varepsilon \sigma_n\right)\right| \leqslant \frac{h^{-1}(n)}{C_0 P} \sum_{t=1}^{P} E \xi_t^2 I\left(|S_P| \geqslant \varepsilon \sigma_n\right) \leqslant$$

$$\leqslant \frac{h^{-1}(n)}{C_0} E \xi_t^2 I\left(|\xi_t| \geqslant \frac{\varepsilon \sigma_n}{2}\right) + \frac{h^{-1}(n)}{P C_0} \sum_{t=1}^{P} E \xi_t^2 I\left(\sum_{\substack{s=1 \\ s \neq t}}^{P} \xi_s \geqslant \frac{\varepsilon \sigma_n}{2}\right) \leqslant$$

$$\leqslant \frac{h^{-1}(n)}{C_0} E \xi_t^2 I\left(|\xi_t| \geqslant \frac{\varepsilon \sigma_n}{2}\right) + \frac{h^{-1}(n)(1+\psi(1))}{C_0 P \sigma_n^2 \varepsilon^2} \sum_{t=1}^{P} E \xi_t^2 E\left(\sum_{\substack{s=1 \\ s \neq t}}^{P} \xi_s\right)^2 =$$

$$= \frac{h^{-1}(n)}{C_0} E \xi_t^2 I\left(|\xi_t| \geqslant \frac{\varepsilon \sigma_n}{2}\right) + \frac{C E \xi_t^2 E\left(\sum_{s=1}^{P} \xi_s\right)^2}{h(n) \sigma_n^2} =$$

$$= \frac{1}{c_0\, h(n)} E\, \xi_t^2\, I\left(|\xi_t| \geqslant \frac{\varepsilon \delta_n}{2}\right) + \frac{C\, h(P)\, P}{h(n)\cdot n} \to 0\,, \qquad n \to \infty.$$

Theorem is proven.

Since according to Corollary 3.1.2 the condition $\psi(1) < 1$ implies $\delta_n^2 \to \infty$, as $n \to \infty$ then all theorems may be re-formulated for example in the following way:

Theorem 5.7.2. Let ξ_t, $t \in \mathbb{Z}^1$ be a centered ψ-mixing s.r.p. such that for some $\delta > 0$, $E|\xi_0|^{2+\delta} < \infty$ and $\psi(1) < 1$. Then for the process the invariance principle holds.

or

Theorem 5.7.3. Let ξ_t, $t \in \mathbb{Z}^1$, be a centered ψ-mixing s.r.p. such that for some $\delta \geqslant 0$, $E|\xi_0|^{2+\delta} < \infty$, $\psi(1) < 1$ and $\psi(n) = O\left(\frac{1}{\ell n^{1+\rho}n}\right)$ as $n \to \infty$. Then for the process the law of iterated logarithm holds.

6. LIMIT THEOREMS UNDER GENERALIZED MIXING CONDITIONS

In /29/ Dobrushin has introduced some weak dependence conditions which represent a natural generalization of the mixing conditions for random fields. He suggested also that the c.l.th. which contains well-known results as a special case under these generalized mixing conditions holds. In this chapter a theorem of such type will be proved for s.r.p. with generalized α -mixing condition. Note that similar results may be obtained both for s.r.p. with generalized φ -mixing condition and for random fields. In the conclusion some generalizations of the c.l. th. will be given in case of so-called non-commutative probability theory.

6.1 Distances in Space of Probability Measures, Kantorovich-Vasershtein Metric

Let X be a complete separable metric space with $\rho(x,\tilde{x})$ a metric, measurable with respect to pair (x , $\tilde{x}$), $x, \tilde{x} \in X$. Denote $\mathfrak{M}$ the space of all probability distributions on $(X, \mathcal{B})$, where $\mathcal{B}$ is the Borel σ -algebra of X . The quantity

$$\mathbb{R}(P, Q) = \inf E \rho(\eta, \varsigma), \qquad (6.1.1.)$$

where inf is taken over all pairs (η, ς) of random variables, such that their marginal distributions coincide with P , Q , respectively, is called Kantorovich-Vasershtein metric (see /32/ /72/,/126/.).

Let $\pi(P, Q)$ be the Levy-Prokhorov metric, defined as the lower boundary of the numbers $\varepsilon > 0$ such that for all closed sets $F \subset X$ the following relations are valid:

$$P(F) - Q(F) \leqslant \varepsilon, \quad Q(F) - P(F_\varepsilon) \leqslant \varepsilon , \qquad (6.1.2)$$

where F_ε is the ε -neighbourhood of the set F .

It is well-known that the topology of weak convergence is metrized by the Levy-Prokhorov metric. The Kantorovich-Vasershtein metric has a similar property, namely, the following statements(the second of which we give without proof) are valid.

Theorem 6.1.1 (Dobrushin). Let P_0 be a probability distribution on $(X, \mathcal{B})$ and $\mathcal{U}(P_0)$ a family of all probability distributions on $(X, \mathcal{B})$, such that $R(P_0, P) < \infty$. Then $\mathcal{U}(P_0)$ with the metric (6.1.1) forms a complete metric space. If the sequence of distributions P_n converges to P in a sense of Kantorovich-Vasershtein metric, then P_n tends to P in the topology of weak convergence.

Proposition 6.1.1. Let $W \subset \mathcal{U}(P_0)$ be such that for some $x_0 \in X$

$$\lim_{c \to \infty} \int_{\{x : \rho(x, x_0) > c\}} \rho(x, x_0)\, P(dx) = 0 \qquad (6.1.3)$$

uniformly with respect to all $P \in W$. Then the metric (6.1.3) is equivalent to the topology of weak convergence on W.

It follows from Proposition 6.1.1 that if the metric $\rho(x, \widetilde{x})$, $x, \widetilde{x} \in X$, is bounded, then the Kantorovich-Vasershtein metric metrizes the topology of weak convergence. Before proving Theorem 6.1.1 let us obtain some preliminary results.

Lemma 6.1.1. Between Levy-Prokhorov and Kantorovich-Vasershtein metrics the following relation is valid:

$$\mathbb{R}(P, Q) \geqslant (\pi(P, Q))^2. \qquad (6.1.4)$$

Proof. Let the vector (η, ς) have P and Q as its marginal distributions, correspondingly, and let Π be its distribution. Then we have

$$\mathbb{R}(P, Q) = \inf \left\{ \iint_{X \times X} \rho(x, y)\, \Pi(dx, dy) \right\}. \qquad (6.1.5)$$

Suppose there exists a close set F such that one of the inequalities (6.1.2) is not fulfilled, for example,

$$Q(F) - P(F_\varepsilon) \geqslant \varepsilon.$$

Hence

$$\iint_{X F} \Pi(dx, dy) \geqslant \iint_{F_\varepsilon X} \Pi(dx, dy) + \varepsilon.$$

Further we have

$$\iint_{F_\varepsilon' F} \Pi\,(dx, dy) = \iint_{\{(x,y):\; \rho(x,y)\geqslant\varepsilon,\; x\in X,\; y\in F\}} \Pi\,(dx, dy) \qquad \geqslant \varepsilon .$$

Finally

$$\frac{1}{\varepsilon}\, E\,\rho\,(\eta, \varsigma) \geqslant \varepsilon$$

which proves Lemma 6.1.1.

Now we represent a compactness criterion of a family of probability measures η , $\eta \subset \mathcal{M}$.

__Lemma 6.1.2.__ For a set of probability distributions η to be compact the following condition is necessary and sufficient: There exists some non-negative function $h(x)$, $x \in X$, (capable of assuming the value $+\infty$) such that for any d , $d \geqslant 0$, $d < \infty$ the set $\{\,x : h(x) \leqslant d,\; x \in X\,\}$ is compact and

$$\int_X h(x)\,P(dx) \leqslant C < \infty \qquad \text{for each } P \in \eta . \tag{6.1.6}$$

__Proof.__ As it was mentioned in Chapter 1, the following condition is necessary and sufficient for the compactness of η : for any $\varepsilon > 0$ there exists a compact set $K_\varepsilon \subset X$ such that $P(K_\varepsilon) \geqslant 1 - \varepsilon$ for all $P \in \eta$. From (6.1.6) it follows

$$P\left(x : h(x) > \frac{C}{\varepsilon}\right) \leqslant \varepsilon .$$

Hence the sufficiency of the conditions of Lemma 6.1.2 is valid. On the other hand if the set η is compact then the required function can be constructed putting $h(x)$ equal to the minimal integer $n \geqslant 0$, for which $x \in K_{2^{-n}}$. If there are no such integers, then we put $h(x) = +\infty$. Lemma 6.1.2 is proven.

__Lemma 6.1.3.__ Let a sequence P_n , $n \in \mathbb{N}$, of probability distributions on X weakly converge to __the distribution__ P . If the function $f(x)$, $x \in X$, (which is able to take the value $+\infty$) is such that for any d , $0 \leqslant d < \infty$, the set $\{\,x : f(x) > d\,\}$ is open, then

$$\lim_{n \to \infty} \int_X f(x)\,P_n(dx) \geqslant \int_X f(x)\,P(dx).$$

The validity of this lemma leads out from the inequality

$$\varliminf_{n \to \infty} P_n\big(\{x : f(x) > d\}\big) \geq P\big(\{x : f(x) > d\}\big), \quad 0 \leq d < \infty.$$

The latter is fulfilled since the set $\{x : f(x) > d\}$ is open (the proof of this fact one can found, for example, in /98/).

<u>Proof of Theorem 6.1.1.</u> First we demonstrate that the quantity (6.1.1) is a metric. From Lemma 6.1.1 and the fact that $\pi(P,Q)$ is a metric it follows that if $\mathbb{R}(P,Q) = 0$ then $P = Q$. Now let us verify the validity of the triangle inequality for $\mathbb{R}(P,Q)$. Suppose that P, Q, S are some probability distributions on $(X, \mathcal{B})$. Let $(\xi, \eta), (\tilde{\eta}, \zeta)$ be two pairs of random vectors such that ξ has the distribution P, ζ has the distribution S as for η and $\tilde{\eta}$ let they be identically Q distributed. Then one can construct the random variables ξ_1, η_1, ζ_1 such that the mutual distributions of vectors $(\tilde{\eta}, \zeta)$ and (η_1, ζ_1) as well as those of vectors (ξ, η) and (ξ_1, η_1) is the same (for example, if ξ_1, η_1, ζ_1 form a Markov chain). Hence

$$\mathbb{R}(P, S) \leq E\rho(\xi_1', \zeta_1') \leq E\rho(\xi_1', \eta_1') + E\rho(\eta_1', \zeta_1'). \qquad (6.1.7)$$

Note for the appropriate choice of the random variables $\xi, \eta, \tilde{\eta}, \zeta$ the right-hand side of (6.1.7) will be arbitrarily close to $\mathbb{R}(P,Q) + \mathbb{R}(Q,S)$. Thus the triangle inequality is proven. From Lemma 6.1.1 it follows immediately that the topology of weak convergence is metrized by Kantorovich-Vasershtein metric. Now let us prove the completeness of space $\mathcal{U}(P_0)$. From Lemma 6.1.1 it follows also that if some sequence P_n is fundamental in Kantorovich-Vasershtein metric, then it is the same in Levy-Prokhorov metric and thus has a weak limit $\overline{P}$. From definition (6.1.1) it is evident that there exist the random variables $\eta_{n,m}$ and $\tilde{\eta}_{n,m}$, $nm \in \mathbb{N}$, with distributions P_n and P_m, respectively, such that the following inequality is valid:

$$E\rho(\eta_{n,m}, \tilde{\eta}_{n,m}) \leq 2\mathbb{R}(P_n, P_m). \qquad (6.1.8)$$

The sequence of distributions P_n, $n \in \mathbb{N}$, is compact and therefore from Lemma 6.1.2 it follows that the family of mutual distributions of $(\eta_{n,m}, \tilde{\eta}_{n,m})$, $n, m \in \mathbb{N}$, is also compact in the topology of weak convergence of space X^2 with the metric

$$\rho_{(2)}(x', x'') = \rho(x'_1, x''_1) + \rho(x'_2, x''_2), \quad x' = (x'_1, x'_2), \quad x'' = (x''_1, x''_2).$$

Let $(\bar\eta_n, \bar\eta)$ be the weak limit of the sequence $(\eta_{n,m_i}, \tilde\eta_{n,m_i})$ for some subsequence $m_i \to \infty$ as $i \to \infty$. Then P_n and $\bar P$ are the distributions of random variables $\bar\eta_n$ and $\bar\eta$, respectively. From (6.1.8) and Lemma 6.1.3 it follows

$$E\rho(\bar\eta_n, \bar\eta) \leqslant 2 \varlimsup_{i \to \infty} \mathbb{R}(P_n, P_{m_i}). \qquad (6.1.9)$$

Since the sequence P_n is fundamental then

$$E\rho(\bar\eta_n, \bar\eta) \to 0, \quad n \to \infty$$

and finally

$$\mathbb{R}(P_n, \bar P) \to 0, \quad n \to \infty.$$

Theorem 6.1.1 is proven.

<u>Theorem 6.1.2</u>. If the metric $\rho(x, \tilde x)$ is discrete, i.e.,

$$\rho(x, \tilde x) = \begin{cases} 1, & x \neq \tilde x \\ 0, & x \neq \tilde x \end{cases}, \qquad x, \tilde x \in X,$$

then

$$\mathbb{R}(P, Q) = \sup_{B \in \mathcal{B}} |P(B) - Q(B)|,$$

namely $\mathbb{R}(P, Q)$ coincides with known variation distance.

<u>Proof</u>. According to the Hahn-Jordan theorem (see, for example, /79/) we have

$$P(B) = \mu(B) + \lambda_1(B), \quad Q(B) = \mu(B) + \lambda_2(B), \quad B \in \mathcal{B},$$

where $\mu, \lambda_1, \lambda_2$ are finite measures and λ_1, λ_2 are concentrated on some non-intersecting sets A and $X \setminus A$, respectively . Moreover,

$$\sup_{B \in \mathcal{B}} |P(B) - Q(B)| = 1 - \mu(X).$$

For any pair of random variables η and ζ with corresponding distributions P and Q we can write

138

$$E\rho(\eta,\zeta)=P\,(\eta\neq\zeta)\geqslant 1-P\,(\eta\in A,\zeta\in A)-$$

$$-P\,(\eta\in X\backslash A,\ \zeta\in X\backslash A)\geqslant 1-P\,(\zeta\in A)-P\,(\eta\in X\backslash A)=$$

$$=1-\mu(A)-\lambda_2(A)-\mu(X\backslash A)-\lambda_1(X\backslash A)=1-\mu(X).$$

On the other hand one can construct the pair of random variables η,ζ of the distributions P,Q respectively and such that

$$P\,(\eta\neq\zeta)=1-\mu(X)$$

putting

$$P(\eta\in A,\zeta\in B)=\mu(A\cap B)+\lambda_1(A)\lambda_2(B)\lambda_1^{-1}(X).$$

Really, since $\quad\lambda_1(X)=\lambda_2(X)=1-\mu(X)\quad$ we have

$$P\,(\eta\in B)=P\,(\eta\in B,\zeta\in X)=\mu(B)+\lambda_1(B),$$

$$P\,(\zeta\in B)=P\,(\eta\in X,\zeta\in B)=\mu(B)+\lambda_2(B)$$

and

$$P\,(\eta\neq\zeta)=\lambda_1^{-1}(x)\int\limits_X\Big(\int\limits_{X\backslash\{x\}}\lambda_2(dy)\Big)\lambda_1(dx)=$$

$$=\lambda_1^{-1}(X)\lambda_1(A)\lambda_2(X\backslash A)=\lambda_2(X)=1-\mu(X).$$

<u>Theorem 6.1.3 (/72/,/125/).</u> Let $X=R^1$ with ordinary Euclidean metric. Then

$$R(P,Q)=\int\limits_{-\infty}^{\infty}|F(x)-G(x)|\,dx\,,\qquad\qquad (6.1.10)$$

where F and G are distribution functions of measures P and Q, respectively.

Before proving Theorem 6.1.3 we give a well-known technical result (see, for example, /7/).

__Lemma 6.1.4.__ Let ζ be a random variable uniformly distributed on $[0,1]$ and F be some distribution function. Define function $F^{-1}(y)$ on $[0,1]$ in the following way:

1) if for $y \in [0,1]$ there exists a number x such that $F(x) = y$ then we put $F^{-1}(y) = x$ (if such x is not unique then we take any of them),

2) if there are no such x, then we put

$$F^{-1}(y) = \inf_{z > y} F^{-1}(z),$$

where inf takes over all such $z > y$, for which $F^{-1}(z)$ is defined in 1),
then the distribution function of the random variable $F^{-1}(\zeta)$ is F.

__Proof of Theorem 6.1.3.__ Let's assume that

$$\int_{-\infty}^{\infty} |F(x) - G(x)|\, dx < \infty$$

and prove that the left-hand side in (6.1.10) does not exceed the right-hand one. Denote

$$\xi = F^{-1}(\zeta), \qquad \eta = G^{-1}(\zeta).$$

Hence

$$E|\xi - \eta| = \int_{0}^{1} |F^{-1}(y) - G^{-1}(y)|\, dy =$$

$$= \int_{-\infty}^{\infty} |F(x) - G(x)|\, dx$$

and

$$\mathbb{R}(P,Q) \leqslant \int_{-\infty}^{\infty} |F(x) - G(x)|\, dx.$$

Now we will show that the inverse inequality also holds. Suppose that $\mathbb{R}(P,Q) < \infty$ and ξ, η are the random variables with distributions P, Q, respectively, such that

$$E|\xi - \eta| < \infty.$$

Let $\alpha = \max\{\xi - \eta, 0\}$, $\beta = \max\{\eta - \xi, 0\}$.

Then

$$E|\xi - \eta| = E\alpha + E\beta.$$

140

Further we have

$$E\beta = E\left(E\left(\beta\,|\,\xi=x\right)\right) \qquad \text{, and}$$

$$E\left(\beta\,|\,\xi=x\right) < \infty \qquad \text{a.s. by measure } P\;.$$

Further

$$E\left(\beta\,|\,\xi=x\right) = \int\limits_{+0}^{\infty} z\,P\left(\beta\in dz\,|\,\xi=x\right) = \int\limits_{+0}^{\infty} P\left(\beta \geqslant z\,|\,\xi=x\right)dz.$$

Hence

$$E\beta = \int\limits_{-\infty}^{\infty} P\left(\xi\in dx\right)\int\limits_{+0}^{\infty} P\left(\eta-\xi \geqslant z\,|\,\xi=x\right)dz =$$

$$= \int\limits_{-\infty}^{\infty} P\left(\xi\in dx\right)\int\limits_{+0}^{\infty} P\left(\eta \geqslant x+z\,|\,\xi=x\right)dz =$$

$$= \int\limits_{-\infty}^{\infty} P\left(\xi\in dx\right)\int\limits_{x+0}^{\infty} P\left(\eta \geqslant y\,|\,\xi=x\right)dy =$$

$$= \iint\limits_{\{y>x\}} P\left(\xi\in dx,\ \eta \geqslant y\right)dy = \int\limits_{-\infty}^{\infty} P\left(\xi<y,\ \eta \geqslant y\right)dy.$$

Here we use the change of variable $z = y - x$. Further we have

$$E\,|\xi-\eta| = \int\limits_{-\infty}^{\infty}\left(P\left(\xi<y,\ \eta \geqslant y\right)+P\left(\xi \geqslant y,\ \eta < y\right)\right)dy =$$

$$= \int\limits_{-\infty}^{\infty}\left(P\left(\xi<y\right)+2\,P\left(\eta<y\right)-2\,P\left(\xi<y,\ \eta<y\right)\right)dy \geqslant$$

$$\geqslant \int\limits_{-\infty}^{\infty}\left[F(y)+G(y)|-2\min\left(F(y),\ G(y)\right)\right]dy =$$

$$= \int\limits_{-\infty}^{\infty}|F(y)-G(y)\,dy.$$

Note Theorem 6.1.6 exactly in the same way may be generalized in the case $X = R^K,\ k>1.$

Let ξ_t, $t \in \mathbb{Z}^1$, be a s.r.p. which assumes values in X and let $\mathbb{P} = \{ P_V, \ V \subset \mathbb{Z}^1 \}$ be the set of its finite-dimensional distributions. Here each P_V is a probability measure on σ-field of Borel subsets of space $\{ X^{|V|}, \rho_V \}$, where

$$X^{|V|} = \{ (x_1, \ldots, x_{|V|}), x_i \in X, \ i = \overline{1, |V|} \}, \tag{6.2.1}$$

$$\rho_V (x, \tilde{x}) = \sum_{t=1}^{|V|} \rho(x_t, \tilde{x}_t), \qquad x, \tilde{x} \in X^{|V|}.$$

We say a s.r.p. ξ_t, $t \in \mathbb{Z}^1$, satisfies the generalized α-mixing condition if for any $k, m, n \in \mathbb{N}$ the following inequality holds:

$$\mathbb{R} \left(P_{I_{-k}^0 \cup I_n^{n+m}}, \ P_{I_{-k}^0} \times P_{I_n^{n+m}} \right) \leq \alpha_\rho(n), \tag{6.2.2}$$

where $\alpha_\rho(n) \to 0$ as $n \to \infty$,

$I_a^b = \{ t \in \mathbb{Z}^1 : 0 < t < b \}$, $a, b \in \mathbb{Z}^1$, and Kantorovich-Vasershtein metric is calculated here with respect to metric (6.2.1). Obviously, if $X = R^1$ and metric ρ is discrete, then (6.2.2) is equivalent to

$$| P(AB) - P(A)P(B)| \leq \alpha(n) \tag{6.2.3}$$

for any $A \in \sigma(\xi_t, t \leq 0)$, $B \in \sigma(\xi_t, t \geq n)$ and $\alpha(n) \to 0$, as $n \to \infty$,
i.e., it coincides with the usual α-mixing condition.
By changing the space X and the metric ρ one can obtain various mixing conditions. For example, if $X = R^1$, $\rho(x, \tilde{x}) = |x - \tilde{x}|$, $x, \tilde{x} \in R^1$, then (6.2.2) reduces to the following:
For any $k, m, n \in \mathbb{N}$

$$\int_{R^{k+m}} | F_{k+m} (x_t, t \in I_{-k}^0 \cup I_n^{n+m}) - \tag{6.2.4}$$

$$- F_k (x_t, t \in I_{-k}^0) F_m (x_t, t \in I_n^{n+m}) \prod_{t \in I_{-k}^0 \cup I_n^{n+m}} dx_t \leq \hat{\alpha}(n),$$

where

$$F_p(x_t,\ t\in I_a^6) = P(\bigcap_{t\in I_a^6}(\xi_t < x_t)),\quad a,6\in\mathbb{Z}^1,\quad a<6,$$

$\hat{\alpha}(n)$ does not depend on k, m and tends to zero as $n\to\infty$. Analogously we can introduce the generalized φ -mixing condition for the processes. Let for any $k,m,n\in\mathbb{N}$, $P_{I_n^m}(\cdot/x_t,\ t\in I_{-k}^0)$ be the conditional distribution for random variables ξ_t, $t\in I_n^m$, under condition that $\xi_t = x_t$, $t\in I_{-k}^0$. We say that s.r.p. $\xi_t,\ t\in\mathbb{Z}^1$ with values in X satisfies the generalized φ -mixing condition if with probability 1 concerning the measure $P_{I_n^{n+m}}$ the following inequality holds:

$$\mathbb{R}\big(P_{I_n^{n+m}}(\cdot/x_t,\ t\in I_{-k}^0),\ P_{I_n^{n+m}}\big) \le \varphi(n),\qquad (6.2.5)$$

where $\varphi(n)\to 0$, as $n\to\infty$ and does not depend on K and m . Note that Kantorovich–Vasershtein metric in (6.2.5) is calculated with respect to the metric:

$$\rho_V(x,\tilde{x}) = \max_{t\in V}\{\rho(x_t,\tilde{x}_t)\},\quad x,\tilde{x}\in X^{|V|}.$$

As in the case of ordinary mixing conditions for the proof of limit theorems it is very important to have **estimates of the covariance between the random variables.**

Lemma 6.2.1. Let a s.r.p. $\xi_t,\ t\in\mathbb{Z}^1$, with values in (X,ρ) satisfy the generalized α-mixing condition and $V_1 = (-k,0)$, $V_2 = (n,n+m)$ be the intervals in $\mathbb{Z}^1$. Let $g_i(x_t,t\in V_i)$ $i = 1,2,$ be the continuous functions with respect to metric ρ_{V_i} $i = 1,2$, respectively, and let $I_{V_i}(\gamma)$, $i = 1,2$, be their continuity moduli. Suppose that for some s , $u>1$, $s^{-1}+u^{-1}\le 1$ the moments

$$E\,|\,g_1(\xi_t,\ t\in V_1)|^s,\quad E\,|\,g_2(\xi_t,\ t\in V_2)|^u$$

exist.
Then for any $\gamma > 0$ the following inequality holds:

$$|E g_1(\xi_t,\ t\in V_1)g_2(\xi_t,\ t\in V_2) - E g_1(\xi_t,\ t\in V_1)E g_2(\xi_t,\ t\in V_2)| \le$$
$$\hspace{10cm}(6.2.6)$$
$$\le \tau_{V_1}(\gamma)E\,|g_2(\xi_t,\ t\in V_2)| + \tau_{V_2}(\gamma)E\,|g_1(\xi_t,\ t\in V_1)| +$$
$$+ BE^{1/s}\,|g_1(\xi_t,t\in V_1)|^s E^{\frac{1}{u}}\,|g_2(\xi_t,\ t\in V_2)|^u \left(\frac{\alpha(n)}{\gamma}\right)^{1-\frac{1}{s}-\frac{1}{u}},$$

$$0 < B < \infty.$$

If with probability 1 $\ |g_i(\xi_t,\ t \in V_i)| \leq C_i < \infty$, $\qquad i = 1,2,$
then the right-hand side of inequality (6.2.6) may be replaced
by

$$\tau_{V_1}(\gamma)\, E\,|\,g_2(\xi_t,\ t \in V_2)| + \tau_{V_2}(\gamma)\, E\,|\,g_1(\xi_t, t \in V_1)| +$$
$$+ \ \frac{B_2 C_1 C_2}{\gamma}\ \alpha(n). \tag{6.2.7}$$

<u>Proof.</u> Denote

$$g(x_t,\ t \in V_1 \cup V_2) = g_1(x_t,\ t \in V_1)\, g_2(x_t,\ t \in V_2),$$
$$\xi_{V_i} = (\xi_t,\ t \in V_i),\quad i = 1,2,$$

and assume that the random vector $\ \eta_{V_1 \cup V_2} = (\eta_t,\ t \in V_1 \cup V_2)$
has the distribution $\ P_{V_1} \times P_{V_2}\ ,\ P_{V_i} \in \mathfrak{M}\ $, i = 1,2.
Let

$$A_\gamma = \left\{ \omega:\ \rho_{V_1 \cup V_2}(\xi_{V_1 \cup V_2},\ \eta_{V_1 \cup V_2}) < \gamma \right\},$$

A_γ' be the complement of $\ A_\gamma\ $ and with probability 1

$$|g_i(\xi_t,\ t \in V_i)| \leq C_i < \infty,\quad i = 1,2.$$

Then we can write

$$|E\, g_1(\xi_{V_1})\, g_2(\xi_{V_2}) - E\, g_1(\xi_{V_1})\, E\, g_2(\xi_{V_2})| =$$

$$= |\,E\, g(\xi_{V_1 \cup V_2}) - E\, g(\eta_{V_1 \cup V_2})| \leq$$

$$\leq E_{A_\gamma}|\,g(\xi_{V_1 \cup V_2}) - g(\eta_{V_1 \cup V_2})| + E_{A_\gamma'}|\,g(\xi_{V_1 \cup V_2}) - g(\eta_{V_1 \cup V_2})| \leq$$

$$\leq E_{A_\gamma}|\,g_1(\xi_{V_1})|\,|\,g_2(\xi_{V_2}) - g_2(\eta_{V_2})| + E_{A_\gamma'}|\,g_2(\eta_{V_2})|\,|\,g_1(\eta_{V_1}) - g_1(\xi_{V_2})| +$$

$$+ \ \frac{B_1 C_1 C_2}{\gamma}\ \alpha(n) \leq \tau_2(\gamma)\, E\,|\,g_1(\xi_{V_1})| + \tau_1(\gamma)\, E\,|\,g_2(\xi_{V_2})| +$$

$$+ \ \gamma^{-1}\, B_1 C_1 C_2\, \alpha(n).$$

Thus inequality (6.2.7) is proven.
The idea of this inequality belongs to Dobrushin (see inequality
(3.8) in his article /32/ ; note that this inequality contains a gap: γ and $\delta(\gamma)$ should be replaced).

Now we prove inequality (6.2.6). Denote

$$g_i^{K_i}(x) = \begin{cases} g_i(x), & |g_i(x)| \le K \\ K_i, & g_i(x) > K_i \\ -K_i, & g_i(x) < -K_i \end{cases}$$

$$K_i \in R_+^1, \qquad \bar{g}_i^{K_i}(x) = g_i(x) - g_i^{K_i}(x), \quad x \in X^{|V_i|}, \quad i = 1,2.$$

Let $\tau_i^{K_i}(\gamma)$ be the continuity moduli of $g_i^{K_i}(x)$.
It is easy to check that

$$\tau_i^{K_i}(\gamma) \le \tau_i(\gamma), \qquad |g_i^{K_i}(x)| \le K_i, \qquad\qquad i = 1,2.$$

Further

$$|E g_1(\xi_{V_1}) g_2(\xi_{V_2}) - E g_1(\xi_{V_1}) E g_2(\xi_{V_2})| \le$$

$$\le E |g_1^{K_1}(\xi_{V_1}) g_2^{K_2}(\xi_{V_2}) - g_1^{K_1}(\eta_{V_1}) g_2^{K_2}(\eta_{V_2})| +$$

$$+ E |g_1^{K_1}(\xi_{V_1}) \bar{g}_2^{K_2}(\xi_{V_2})| + E |\bar{g}_1^{K_1}(\xi_{V_1}) g_2^{K_2}(\xi_{V_2})| +$$

$$+ E |\bar{g}_1^{K_1}(\xi_{V_1}) \bar{g}^{K_2}(\xi_{V_2})| + E |g_1^{K_1}(\xi_{V_1})| E |\bar{g}_2^{K_2}(\xi_{V_2})| +$$

$$+ E |\bar{g}_1^{K_1}(\xi_{V_1})| E |g_2^{K_2}(\xi_{V_2})| + E |\bar{g}_1^{K_1}(\xi_{V_1})| E |\bar{g}_2^{K_2}(\xi_{V_2})|.$$

Now for concluding our proof it is sufficient to put

$$K_1 = \left(\frac{\gamma E |g_1(\xi_{V_1})|^s}{\alpha(n)} \right)^{1/s},$$

$$K_2 = \left(\frac{\gamma E |g_2(\xi_{V_2})|^u}{\alpha(n)} \right)^{1/u}$$

and to proceed it in the same way as in /32/ .

Let $f(x)$, $x \in X$, be a continuous function defined on X with continuity moduli $\tau(\gamma)$, $\gamma \in R_+^1$, i.e.,

$$\tau(\gamma) = \sup_{(x,\tilde{x}):\ \rho(x,\tilde{x}) \leqslant \gamma} |f(x) - f(\tilde{x})|.$$

We will say that a random process ξ_t, $t \in Z^1 (\xi_t \in X)$, satisfies the c.l.th. with function f if for any $s \in R^1$ the following relation holds:

$$\lim_{n \to \infty} P\left(\left(D \sum_{t=1}^{n} f(\xi_t) \right)^{-\frac{1}{2}} \sum_{t=1}^{n} \left(f(\xi_t) - E f(\xi_t) \right) < s \right) =$$

$$= \frac{1}{\sqrt{2\pi}} \int_{-\infty}^{s} e^{-u^2/2}\, du.$$

__Theorem 6.3.1.__ Let $\xi_t, t \in Z^1$, be a generalized α -mixing s.r.p. with values in (X, ρ) and the following conditions are valid:

1) The function $f(x)$, $x \in X$ is such that for some $\delta > 0$

$$E |f(\xi_t)|^{2+\delta} < \infty, \quad (\quad |f(\xi_t)| < C \quad \text{with probability 1}).$$

2) There exists a sequence of numbers γ_n, $n = 1, 2 \ldots,$ decreasing to zero such that:

$$\gamma_n^{-1} \alpha_\rho (n) \leqslant \beta(n), \quad \beta(n) \downarrow 0 \quad \text{as} \quad n \to \infty$$

and

$$\sum_{n=1}^{\infty} \tau(\gamma_n) < \infty, \quad \sum_{n=1}^{\infty} \beta^{\frac{\delta}{2+\delta}}(n) < \infty \quad \left(\quad \sum_{n=1}^{\infty} \beta(n) < \infty \right).$$

Then the series

$$\sigma_f^2 = E f^2(\xi_1) + 2 \sum_{n=2}^{\infty} \left(E f(\xi_1) f(\xi_n) - E f(\xi_1) E f(\xi_n) \right)$$

converges and if $\sigma_f^2 > 0$, the process ξ_t, $t \in Z^1$, satisfies c.l.th. with function f.

Note that if $X = R^1$ and ρ is the discrete metric, then the continuity modulus $\tau^f(\gamma)$, $\gamma \in R_+^1$ of any function $f(x)$, $x \in R^1$ is equal to zero for $\gamma < 1$ .Hence the Ibragimov theorem (see Chapter 5)follows from Theorem 6.3.1. If $X = R^1$, $\rho(x,\tilde{x}) = |x - \tilde{x}|$, $x, \tilde{x} \in R^1$, $f(x) = x$, then $\tau^f(\gamma) = \gamma$, $\gamma \in R_+^1$, and the following results are obtained from the same theorem.

__Theorem 6.3.2.__   Suppose that a s.r.p. ξ_t, $t \in \mathbb{Z}^1$, with values in R^1 satisfies the mixing condition (6.2.4) and for some $\delta > 0$, $E|\xi_t|^{2+\delta} < \infty$. If for some $\varepsilon > 0$ the series

$$\sum_{n=1}^{\infty} n^{1+\varepsilon} \left(\hat{\alpha}(n) \right)^{\delta/2 + \delta}$$

converges, then

$$\sigma^2 = E\left(\xi_1 - E\xi_1\right)^2 + 2 \sum_{n=2}^{\infty} E\left(\xi_1 - E\xi_1\right)\left(\xi_n - E\xi_n\right) < \infty .$$

If in addition $\sigma^2 > 0$, then the process satisfies c.l.th. with function $f(x) = x$, $x \in R^1$.

__Proof of Theorem 6.3.1.__ It is sufficient to verify the conditions of Lemma 4.1.3 for the process

$$\eta_t = f(\xi_t), \quad t \in \mathbb{Z}^1 .$$

We have

$$D\left(\sum_{t=1}^{n} f(\xi_t) \right) = \sum_{t,s=1}^{n} \left(E f(\xi_t) f(\xi_s) - E f(\xi_t) E f(\xi_s) \right) =$$

$$= n E\left(f(\xi_1) - E f(\xi_1) \right)^2 + 2 \sum_{t=2}^{n} (n-t+1)\left(E f(\xi_1) f(\xi_t) - E f(\xi_1) E f(\xi_t) \right) .$$

Further

$$\lim_{n \to \infty} n^{-1} D\left(\sum_{t=1}^{n} f(\xi_t) \right) = E\left(f(\xi_1) - E f(\xi_1) \right)^2 +$$

$$+ 2 \lim_{n \to \infty} \sum_{t=2}^{n} \left(E f(\xi_1) f(\xi_t) - E f(\xi_1) E f(\xi_t) \right) -$$

$$- 2 \lim_{n \to \infty} n^{-1} \sum_{t=1}^{n} t \left(E f(\xi_1) f(\xi_t) - E f(\xi_1) E f(\xi_t) \right) . \tag{6.3.1}$$

Using Lemma 6.2.1 we can write

$$\left| E f(\xi_1) f(\xi_t) - E f(\xi_1) E f(\xi_t) \right| \leq 2 \; \tau(\gamma_t) + c \left(\frac{\alpha_\rho(t)}{\gamma_t} \right)^{\delta/2 + \delta} ,$$
$$0 < C < \infty .$$

Hence

$$\sigma_f^2 = E\left(f(\xi_1) - E f(\xi_1) \right)^2 + 2 \sum_{t=2}^{\infty} \left(E f(\xi_1) f(\xi_t) - E f(\xi_1) E f(\xi_t) \right) \leq$$

$$\leq E\left(f(\xi_1) - E f(\xi_1) \right)^2 + 2 c \sum_{t=1}^{\infty} \tau(\gamma_t) + 2 c \sum_{t=1}^{\infty} \left(\beta(t) \right)^{\delta/2 + \delta} < \infty ,$$

$$0 < c < \infty .$$

The latter summand in (6.3.1) tends to zero as $n \to \infty$ by Kronecker's lemma. Thus for any $I \subset \mathbb{Z}^1$, $|I| < \infty$, we have

$$DS_I \sim \sigma_f^2 \, |I|, \qquad |I| \to \infty.$$

It remains to check that $\mathfrak{m}_n^{(t)} \to 0$, as $n \to \infty$.
Let

$$W_t(x) = e^{it B \sum_{s=1}^{m} f(x_s)} - 1, \quad m \in \mathbb{N}, \; 0 < B < \infty, \; x \in X^m,$$

where X^m is a metric space with metric (6.2.1).
By means of the following inequality

$$|W_t(x) - W_t(\tilde{x})| \leqslant B|t| \sum_{s=1}^{m} |f(x_s) - f(\tilde{x}_s)|, \quad x, \tilde{x} \in X^m,$$

we conclude that the continuity modulus of function $W_t(x)$ do not exceed $B|t| m \tau^f(\gamma)$, where $\tau^f(\gamma)$ is the continuity modulus of f. Using Lemma 6.2.1 for $j > l$ and $s = 2+\delta$, $u = 2+\delta, \delta > 0$ we have

$$\left| E\, (e^{it S(n,l)} - 1)(e^{it S(n,j)} - 1) \prod_{s=j+1}^{K} e^{it S(n,s)} - \right.$$

$$\left. - E(e^{it S(n,l)} - 1)\, E(e^{it S(n,j)} - 1) \prod_{s=j+1}^{K} e^{it S(n,s)} \right| \leqslant$$

$$\leqslant B_1 \frac{|t|}{\sqrt{n}} \tau(\gamma_{(j-l)q})\, E\,|e^{it S(n,l)} - 1| +$$

$$+ B_2 E^{2/2+\delta} |e^{it S(n,l)} - 1|^{2+\delta} \left(\frac{\alpha_\rho((j-l)q)}{\gamma_{(j-l)q}} \right)^{\delta/2+\delta} \leqslant$$

$$\leqslant B_3 \left(\frac{|t| P \sqrt{P}}{n} \tau(\gamma_{(j-l)q}) + \frac{P^2}{2} (\gamma_{(j-l)q})^{-\delta/2+\delta} \right) (\alpha_\rho (j-l)q)^{\delta/2+\delta}.$$

Let us remind that here

$$S(n,j) = (DS_n)^{-1/2} S_P^{(j)}, \qquad S_P^{(j)} = \sum_{s=(j-1)p+(j-1)q+1}^{jp+(j-1)q} \eta_s,$$

$j = \overline{1, k}$, $k = k(n) = \left[\frac{n}{p+q}\right]$, (p, q) is the standard pair of functions.

Using Lemma 6.2.1 we get

$$\left| E \prod_{j=1}^{K} e^{it S(n,j)} - \prod_{j=1}^{K} E\, e^{it S(n,j)} \right| \leqslant$$

$$\leq B_4 \left(|t|\, p\, n^{-1/2} \sum_{j=1}^{[\frac{n}{p}]} \tau(\gamma_{jq}) + p \sum_{j=1}^{\infty} \left(\frac{\alpha_p(jq)}{\gamma_{jq}} \right)^{\delta/2+\delta} \right).$$

Then

$$\left| E \prod_{j=1}^{K} e^{itS(n,j)} - \prod_{j=1}^{K} E\, e^{itS(n,j)} \right| \leq$$

$$\leq B_4 \left(|t|\, \sqrt{P} \sum_{j=1}^{\infty} \tau^{f}(\gamma_{jq}) + p \sum_{j=1}^{\infty} \beta^{\delta/2+\delta}(jq) \right).$$

By the monotonicity of the members of these series we have

$$\tau(\gamma_{jq}) \leq \frac{2}{q} \sum_{K \geq (j-\frac{1}{2})q}^{jq} \tau(\gamma_K),$$

$$\beta^{\delta/2+\delta}(jq) \leq \frac{2}{q} \sum_{K \geq (j-\frac{1}{2})q}^{jq} \beta^{\delta/2+\delta}(K), \quad j = 1, 2, \ldots$$

Hence

$$\sum_{j=1}^{\infty} \tau(\gamma_{jq}) \leq \frac{2}{q} \sum_{j \geq q/2}^{\infty} \tau(\gamma_j),$$

$$\sum_{j=1}^{\infty} \beta^{\delta/2+\delta}(jq) \leq \frac{2}{q} \sum_{j \geq q/2}^{\infty} \beta^{\delta/2+\delta}(j).$$

Finally

$$\left| E \prod_{j=1}^{K} e^{itS(n,j)} - \prod_{j=1}^{K} E\, e^{itS(n,j)} \right| \leq$$

$$\leq B_4 |t| \frac{2\sqrt{P}}{q} \sum_{j \geq q/2}^{\infty} \tau(\gamma_j) + \frac{2P}{q} \sum_{j \geq q/2}^{\infty} \beta^{\delta/2+\delta}(j). \tag{6.3.2}$$

Now it is evident that there exists a function $q(n)$, $q(n) = o(p)$, $n \to \infty$ such that the right-hand side of (6.2.2) tends to zero as $n \to \infty$.

The following problem concerning the c.l.th. for linear operators (first suggested by Sinai /1/) arose in quantum statistical mechanics in connection with its mathematical demands. In mathematical formalizm of quantum statistical mechanics the observed quantities are identified with self-adjoined linear operators, functioning in some Hilbert space and the calculation of the average values of observed quantities is realized by means of some special functional (state). Here we will not give the exact definitions of the known concepts of quantum statistical mechanics, which may be found, for example, in monographs /16/,/115/.

Let $\mathcal{U}$ be a Von Neumann algebra, ω be the exact normal state on $\mathcal{U}$ and $\mathcal{U}(M)$ be a Von Neumann algebra, generated by M. Denote $\|\cdot\|_\infty$ a norm in $\mathcal{U}$. Let $A \in \mathcal{U}$ be some self-adjoined linear operator and E_λ, $\lambda \in R^1$, be its resolution of the identity. Further for convenience we denote the projector E_λ as $\{A < \lambda\}$. We will call the function $F(\lambda) = \omega(A < \lambda)$ as a distribution function of operator A . The function $\varphi(t) = \omega(\exp(itA))$ will be called characteristic function of operator A . We say that a sequence of self-adjoined operators x_j, $j \in \mathbb{Z}^1$, is stationary, if there exists an automorphism θ of algebra $\mathcal{U}$ preserving the state ω , and such that $x_{j+1} = \theta(x_j)$, $j \in \mathbb{Z}^1$.

Further without losing generality we may consider the stationary sequences of self-adjoined operators such that

$$\omega(x_j) = 0, \qquad \|x_j\|_\infty < 1 .$$

We say that the sequence of operators x_j, $j \in \mathbb{Z}^1$, satisfies an α -mixing condition if:

$$|\omega(PQ) - \omega(P)\omega(Q)| \leq \alpha(n)\|P\|_\infty\|Q\|_\infty$$

for any $P \in \mathcal{U}((x_j)^0_{-\infty})$, $Q \in \mathcal{U}((x_j)^{+\infty}_n)$

and $\alpha(n) \downarrow 0$, $n \to \infty$.

We say that the sequence x_j, $j \in \mathbb{Z}^1$, satisfies φ -mixing condition if:

$$|\omega(PQ) - \omega(P)\omega(Q)| \leq \varphi(n)\omega(P)$$

for any $P \in \mathcal{U}((x_j)^0_{-\infty})$, $Q \in \mathcal{U}((x_j)^{+\infty}_n)$

and $\varphi(n) \downarrow 0$, as $n \to \infty$.

It is said that a sequence of operators $X_j, j \in \mathbb{Z}^1$, is asymptotically commutative if there exists some function $\beta(n)$ decreasing to zero as $n \to \infty$ and such that

$$\| [X_j, X_{j+n}] \|_\infty \leq \beta(n), \qquad j \in \mathbb{Z}^1,$$

where $[X, Y] = XY - YX$ is a commutator.
It is said that a sequence of operators $X_j, j \in \mathbb{Z}^1$, satisfies the c.l.th. if for any fixed $t \in R^1$

$$\omega \left(\sum_{j=1}^{n} X_j < t \, \sigma_n \right) \to \Phi(t) \qquad \text{as} \quad n \to \infty,$$

where

$$\sigma_n^2 = \omega \left(\sum X_j \right)^2.$$

For considered sequences of operators the analogies of Theorem 5.1.4 and **5.1.5** are valid.

<u>Theorem 6.4.1</u> (/115/). Let a stationary α-mixing sequence of **operators** $X_j, j \in \mathbb{Z}^1$, **satisfy** the following conditions:

$$\sum_{n=1}^{\infty} \alpha(n) < \infty, \qquad \sum_{n=1}^{\infty} \beta(n) < \infty.$$

Then

$$\sum_{j=1}^{\infty} | \omega(X_0 X_j + X_j X_0)| < \infty.$$

If in addition

$$\sigma^2 = \omega(X_0^2) + \sum_{j=1}^{\infty} \omega(X_0 X_j + X_j X_0) \neq 0$$

then for the sequence $X_j, j \in \mathbb{Z}^1$, the c.l.th. holds.

<u>Theorem 6.4.2</u> (/46/). Let $X_j, j \in \mathbb{Z}^1$, be a stationary φ-mixing sequence of operators and the following conditions are fulfilled:

1) $\sigma_n^2 \to \infty$, as $n \to \infty$,

2) for some $\varepsilon > 0$, $\sum_{n=1}^{\infty} n^\varepsilon \beta(n) < \infty$,

3) there exists a constant L such that for any projectors

$$P \in U(X_j, j \leq 0) \quad \text{and} \quad Q \in U(X_j, j \geq n)$$

$$|\omega(PQ)| \leqslant L\,\omega(Q) \qquad \text{for} \qquad n \geqslant N_0 > 0.$$

Then for the sequence X_j, $j \in \mathbb{Z}^1$, the c.l.th. holds. In the proofs of these theorems the next so-called Trotter formula (see /124/) plays an important role.

$$\lim_{n \to \infty} \| \exp(i(X+Y)) - (\exp(in^{-1}X)\exp(in^{-1}Y))^n \|_\infty = 0,$$

where X, Y are some self-adjoined operators. With the help of this formula one can obtain the following estimates.

<u>Lemma 6.4.1</u>. For any self-adjoined operators $X, Y \in \mathcal{U}$ the next relation holds:

$$\| \exp(i(X+Y)) - \exp(iX)\exp(iY) \|_\infty \leqslant \tfrac{1}{2} \| [X, Y] \|_\infty .$$

<u>Lemma 6.4.2</u>. Suppose that the stationary sequence of linear operators X_j, $j \in \mathbb{Z}^1$, satisfies φ -mixing condition and there exists a constant L such that for any projectors $P \in \mathcal{U}(X_j,\ j \leqslant 0)$ and $Q \in \mathcal{U}(X_j,\ j \geqslant n)$ the following inequality is valid:

$$|\omega(PQ)| \leqslant L\,\omega(Q).$$

Then for some self-adjoined operators $X \in \mathcal{U}(X_j, j \leqslant 0)$ and $Y \in \mathcal{U}(X_j, j \geqslant n)$ we have

$$|\omega(XY) - \omega(X)\omega(Y)| \leqslant C\,\varphi^{\frac{1}{p}}(n)\|X\|_p \|Y\|_q, \quad 0 < C < \infty,$$

where

$$\|X\|_p = \omega^{\frac{1}{p}}(X^*X)^{p/2}, \qquad \|Y\|_q = \omega^{\frac{1}{q}}(X^*X)^{q/2},$$

$$q > 1, \qquad \tfrac{1}{p} + \tfrac{1}{q} = 1 .$$

Applying these lemmas the proofs of Theorems 6.4.1 and 6.4.2 could be obtained in the usual way with the help of Bernstein's method.

7. LIMIT THEOREMS FOR RANDOM FIELDS

Limit theorems for random fields have their specific properties
caused by the parameter dimensionality. As it was mentioned in
Chapter 2 one should be careful to a certain degree while intro-
ducing the weak dependence conditions for random field, since
it is possible that the random fields classes determined by
these conditions practically do not differ from the set of independent random variables. Another difference from the theory
of limit theorems for the s.r.p. consists in the wide choice
of the passage to infinite volume. We will be interested in the
random fields satisfying the conditions of α- and φ-mixing
in a sense of Chapter 2. As for the applications to statistical
physics let's note that the limit theorems for the random fields
with φ-mixing condition turned out to be interesting, since
Gibbs random fields possess this property.

7.1 Sequences of Sets Tending to Infinity

Let

$$\Lambda(s) = \left\{ t \in \mathbb{Z}^{\nu} : \ 0 \leqslant t^{(i)} \leqslant s^{(i)}, \quad i = \overline{1, \nu} \right\}, \quad s \in \mathbb{Z}^{\nu}_{+},$$

be a ν-dimensional parallelepiped.
It is said that the sequence of parallelepipeds $\Lambda(s_n)$, $s_n \in \mathbb{Z}^{\nu}_{+}$,
tends to infinity as $n \to \infty$ if

$$\min_{1 \leqslant i \leqslant \nu} s_n^{(i)} \to \infty, \qquad n \to \infty .$$

We may define the tendency to infinity for the sets of more general structure as well. The definition given by Van Hove
is mostly used.
Let's denote Λ_t, $t \in \mathbb{Z}^{\nu}$, a shift of the parallelepiped
$\Lambda(s)$ to the vector

$$t s = \left(t^{(1)} s^{(1)}, \ldots, t^{(\nu)} s^{(\nu)} \right),$$

i.e.,

$$\Lambda_t = \Lambda(s) + t s . \tag{7.1.1.}$$

The family of sets $\left\{ \Lambda_t, t \in \mathbb{Z}^{\nu} \right\}$ forms a partitioning of the
lattice $\mathbb{Z}^{\nu}$. For each $I \subset \mathbb{Z}^{\nu}$ let $N^{+}_s(I)$ denote a
number of sets Λ_t such that $\Lambda_t \cap I \neq \emptyset$ and $N^{-}_s(I)$ be
a number of sets Λ_t such that $\Lambda_t \subset I$. It is said that
a sequence of sets $I^{(n)}$ tends to infinity as $n \to \infty$ in a

sense of Van Hove if for any $s \in \mathbb{Z}^\nu$

$$\lim_{n \to \infty} N_s^-(I^{(n)}) = +\infty, \quad \lim_{n \to \infty} \frac{N_s^-(I^{(n)})}{N_s^+(I^{(n)})} = 1. \qquad (7.1.2)$$

An important property of this definition is its invariance with respect to linear transformation of the lattice $\mathbb{Z}^\nu$. Let for any $I \subset \mathbb{Z}^\nu$

$$I_h = \{t \in \mathbb{Z}^\nu : \tau(t, I) \leq h\}, \quad \tau(t, s) = \max_{1 \leq i \leq \nu} |t^{(i)} - s^{(i)}|,$$

$$\tau(t, I) = \inf_{s \in I} \{\tau(t, s)\}.$$

Then the Van Hove definition is equivalent to the following

$$\lim_{n \to \infty} |I^{(n)}| = +\infty, \quad \lim_{n \to \infty} |I^{(n)}|^{-1} |I_h^{(n)}| = 0.$$

One may also use a slightly restrictive Fisher's definition of the sets tendency to infinity. It is said that a sequence of sets tends to infinity as $n \to \infty$ in a sense of Fisher if

$$\lim_{n \to \infty} |I^{(n)}| = \infty$$

and there exists the function $f(h)$, $f(h) \to 0$ as $h \to 0$ such that

$$\sup_n |I^{(n)}|^{-1} |I_{hd(I^{(n)})}^{(n)}| \leq f(h),$$

where $d(I)$ is a diameter of the set I .

This definition is also invariant concerning the linear transformation of the lattice $\mathbb{Z}^\nu$. Let's note that for the sequence of parallelepipeds Fisher's definition is equivalent to the condition: there exists a constant $C > 0$ such that

$$|I^{(n)}| \geq C d^\nu(I^{(n)}).$$

Let's also note that the sequence of parallelepipeds with the different orders to tendency of the sides' lenghth to infinity tends to infinity in sense of Van Hove but not of Fisher.

7.2 Central Limit Theorem for Random Fields

We say that a random field ξ_t , $t \in \mathbb{Z}^\nu$, satisfies the c.l.th. for a sequence of subsets I_n, $I_n \subset \mathbb{Z}^\nu$, $n \in \mathbb{N}$, if

$$\lim_{n\to\infty} P\left(\frac{S_{I_n}-ES_{I_n}}{\sqrt{DS_{I_n}}} < x\right) = \frac{1}{\sqrt{2\pi}}\int_{-\infty}^{x} e^{-\frac{u^2}{2}}du.$$

Below we will establish the next theorems.

Theorem 7.2.1 (/90/). Suppose ξ_t, $t\in\mathbb{Z}^{\nu}$, is a s.r.f. satisfying α -mixing condition and

1. $E\xi_t = 0$, $E|\xi_t|^{2+\delta} < \infty$, $\delta > 0$, $t\in\mathbb{Z}^{\nu}$,

2. $\alpha_{m,n}(\imath) \leqslant f(m)\, n^{\tau}\alpha(\imath)$, $\tau > 0$,

f - is some non-negative function;

3. $\displaystyle\sum_{\imath=1}^{\infty} \imath^{\nu-1}\alpha_{1,1}^{\frac{\delta}{2+\delta}}(\imath) < \infty$, $\alpha(\imath)=0\left(\imath^{(2\tau+1)\nu}\right)$, $\tau > 0$.

Then the series

$$\sigma^2 = \sum_{t\in\mathbb{Z}^{\nu}} E\xi_0\xi_t$$

converges and if $\sigma^2 > 0$ then the random field satisfies the c.l.th. for the sequence of ν -dimensional cubes $I_n=[-n,n]^{\nu}$.

Theorem 7.2.2. Suppose ξ_t, $t\in\mathbb{Z}^{\nu}$, is a s.r.f. satisfying φ -mixing condition and

1. $E\xi_t = 0$, $E\xi_t^2 < \infty$, $t\in\mathbb{Z}^{\nu}$,

2. $\varphi_{m,\infty}(\imath) \leqslant f(m)\,\varphi(\imath)$, where f is some non-negative function,

3. $\displaystyle\sum_{\imath=1}^{\infty} \imath^{\nu-1}\varphi^{\frac{1}{2}}(\imath) < \infty$.

Then the series

$$\sigma^2 = \sum_{t\in\mathbb{Z}^{\nu}} E\xi_0\xi_t$$

converges and if $\sigma^2 > 0$ then the random field satisfies the c.l.th. for the sequence of ν -dimensional cubes $I_n=[-n,n]^{\nu}$.

Proof of Theorem 7.2.1.
As it was shown in Chapter 5 (see the classical proof of Theorem 5.1.4) the condition

$$\sum_{\iota=1}^{\infty} \iota^{\nu-1} \alpha_{1,1}^{\frac{\delta}{2+\delta}}(\iota) < \infty$$

guarantees the convergence of the series $\quad \sigma^2 \quad$, the standard behaviour of the variance, i.e.,

$$DS_{I_n} \sim \sigma^2 |I_n|, \qquad n \to \infty$$

and it also allows us to reduce the proof of the c.l.th. to the random field with bounded components by means of the truncated method. Whence it follows that Theorem 7.2.1 will be proven if in the assumption

$$|\xi_t| < C \qquad \text{with probability 1, } 0 < C < \infty, \ t \in \mathbb{Z}^\nu,$$

we prove the validity of Lemma 4.1.3. As it was mentioned in Section 4.1 in the case of bounded random field components

$$\mathcal{L}_K(\varepsilon) \to 0 \qquad \text{as} \qquad n \to \infty$$

if the function $\quad p(n) \quad$ has the following asymptotics

$$p(n) = O(n^{\frac{1}{2}}) \qquad \text{as} \qquad n \to \infty .$$

Therefore while verifying the conditions of Lemma 4.1.3 we may use the standard pair of functions $\ (p,q)$, where $p(n) = O(n^{\frac{1}{2}})$.

Further let's note that the peculiarity of the considered mixing conditions demands the usage of inequality (3.3.7) in an implicit way.

First applying inequality (3.3.5) we obtain

$$\mathfrak{M}_K \leq \sum_{j=1}^{K-1} | E\, e^{it\hat{S}_{\Delta_j^{(n)}}} \prod_{m=j+1}^{K} e^{it\hat{S}_{\Delta_j^{(n)}}} - E\, e^{it\hat{S}_{\Delta_j^{(n)}}} E \prod_{m=j+1}^{K} e^{it\hat{S}_{\Delta_j^{(n)}}} | \leq$$

$$\leq \sum_{j=1}^{K-1} | E(e^{it\hat{S}_{\Delta_j^{(n)}}} - 1 - it\hat{S}_{\Delta_j^{(n)}} + \frac{t^2}{2}\hat{S}_{\Delta_j^{(n)}} \prod_{m=j+1}^{K} e^{it\hat{S}_{\Delta_m^{(n)}}} -$$

$$- E(e^{it\hat{S}_{\Delta_j^{(n)}}} - 1 - it\hat{S}_{\Delta_j^{(n)}} + \frac{t^2}{2}\hat{S}_{\Delta_j^{(n)}}) E \prod_{m=j+1}^{K} e^{it\hat{S}_{\Delta_m^{(n)}}} | +$$

$$\hspace{10cm} (7.2.1)$$

$$+ |t| \sum_{j=1}^{K-1} | E\hat{S}_{\Delta_j^{(n)}} \prod_{m=j+1}^{K} e^{it\hat{S}_{\Delta_m^{(n)}}} | +$$

$$\frac{t^2}{2} \sum_{j=1}^{K-1} \left| E(\hat{S}_{\Delta_j^{(n)}})^2 \prod_{m=j+1}^{K} e^{it\hat{S}_{\Delta_m^{(n)}}} - E(\hat{S}_{\Delta_j^{(n)}})^2 E\prod_{m=j+1}^{K} e^{it\hat{S}_{\Delta_m^{(n)}}}\right|.$$

Let us show that each of the three sums on the right-hand side of (7.2.1) tends to zero as $n \to \infty$ (further we will denote them T_1, T_2, T_3, respectively). We have

$$T_1 \le 2\frac{|t|^3}{3!} \sum_{j=1}^{K-1} E|\hat{S}_{\Delta_j^{(n)}}|^3 \le C_1 |t|^3 \left(\frac{n}{p}\right)^\nu \frac{E|S_{\Delta_0^{(n)}}|^3}{\sigma_n^3} \le$$

$$\le C_2 |t|^3 \left(\frac{n}{p}\right)^\nu \frac{p^\nu E|S_{\Delta_0^{(n)}}|^2}{(n^\nu)^{\frac{3}{2}}} \le C_2 |t|^3 \left(\frac{p}{n^{\frac{1}{2}}}\right)^\nu \to 0, \quad n \to \infty.$$

Further

$$T_2 \le \sigma_n^{-1}|t| \sum_{j=1}^{K-1} \sum_{s\in\Delta_j^{(n)}} \left| E\xi_s \prod_{m=j+1}^{K} e^{it\hat{S}_{\Delta_j^{(n)}}} - E\xi_s E\prod_{m=j+1}^{K} e^{it\hat{S}_{\Delta_j^{(n)}}}\right| \le$$

$$\le |t|\sigma_n^{-1} \sum_{j=1}^{K-1} \sum_{s\in\Delta_j^{(n)}} \sum_{m=j+1}^{K} \left| E\xi_s\left(e^{it\hat{S}_{\Delta_m^{(n)}}}-1\right)\prod_{l=m+1}^{K} e^{it\hat{S}_{\Delta_l^{(n)}}} - \right.$$

$$\left. - E\xi_s E\left(e^{it\hat{S}_{\Delta_m^{(n)}}}-1\right)\prod_{l=m+1}^{K} e^{it\hat{S}_{\Delta_l^{(n)}}}\right|. \tag{7.2.2}$$

We can make sure in the validity of this step with the help of considerations used in the proof of inequality (3.3.7).
Since $|\xi_t| < C$, $t\in\mathbb{Z}^\nu$, and $|\exp\{it\hat{S}_{\Delta_m^{(n)}}\} - 1| \le C\sigma_n^{-1}|t|p^\nu$ with probability 1 then using Corollary 2.1.1 in (7.2.2) we obtain

$$T_2 \le C_1\sigma_n^{-2}p^\nu|t|^2 \sum_{j=1}^{K-1}\sum_{t\in\Delta_j^{(n)}}\sum_{m=j+1}^{K} \alpha_{1,n^\nu}\left(l(\Delta_j^{(n)},\Delta_m^{(n)})\right) \le$$

$$\le C_2\sigma_n^{-2}p^{2\nu}f(1)n^{\tau\nu}|t|^2\sum_{j=1}^{K-1}\sum_{m=j+1}^{K} \alpha\left(l(\Delta_j^{(n)},\Delta_m^{(n)})\right) \le$$

$$\le C_3 n^{-\nu}p^{2\nu}n^{\tau\nu}|t|^2\left(\frac{n}{p}\right)^\nu\sum_{j=1}^{K} \alpha\left(l(\Delta_0^{(n)},\Delta_j^{(n)})\right) \le$$

$$\leq C_4\, p^{\nu} n^{\tau\nu} |t|^2 \sum_{S=1}^{\infty} \sum_{j:(S-1)q \leq \tau(\Delta_0^{(n)}, \Delta_j^{(n)}) \leq Sq} \alpha\big(\tau(\Delta_0^{(n)}, \Delta_j^{(n)})\big) \leq$$

$$\leq C_5\, |t|^2 p^{\nu} n^{\tau\nu} \sum_{S=1}^{\infty} S^{\nu-1} \alpha(Sq).$$

According to Condition 3) of Theorem 7.2.1 we have

$$T_2 \leq C_5\, \frac{p^{\nu} n^{\tau\nu} |t|^2 \beta(q)}{q^{(2\tau+1)\nu}}, \qquad \beta(n) \to 0, \quad n \to \infty. \tag{7.2.3}$$

Whence it is evident that whatever the function $p = p(n)$ is $p(n) = 0(n^{1/2})$, $p(n) \to \infty$, $n \to \infty$ there will always be found the function $q = q(n) \to \infty$, $q = 0(p)$, $n \to \infty$ such that the right-hand side in 7.2.3 tends to zero as $n \to \infty$.

It remains to estimate the last sum in (7.2.1). We have

$$T_3 \leq \sigma_n^{-2} \frac{t^2}{2} \sum_{j=1}^{K-1} \sum_{t,e \in \Delta_j^{(n)}} \Big| E\,\xi_t \xi_e \prod_{m=j+1}^{K} e^{it\hat{S}_{\Delta_m^{(n)}}} -$$

$$- E\,\xi_t \xi_e\, E \prod_{m=j+1}^{K} e^{it\hat{S}_{\Delta_m^{(n)}}} \Big| \leq$$

$$\leq \sigma_n^{-2} \frac{t^2}{2} \sum_{j=1}^{K-1} \sum_{m=j+1}^{K} \sum_{t,e \in \Delta_j^{(n)}} \Big| E\,\xi_t \xi_e \big(e^{it\hat{S}_{\Delta_{j+1}^{(n)}}} - 1\big) \prod_{m=j+2}^{K} e^{it\hat{S}_{\Delta_m^{(n)}}} -$$

$$- E\,\xi_t \xi_e\, E\big(e^{it\hat{S}_{\Delta_{j+1}^{(n)}}} - 1\big) \prod_{m=j+2}^{K} e^{it\hat{S}_{\Delta_m^{(n)}}} \Big| \leq$$

$$\leq C_1\, |t|^3 \frac{(p^{\nu})^2}{n^{\nu}} \cdot \frac{p^{\nu}}{n^{1/2}} \sum_{j=1}^{K-1} \sum_{m=j+1}^{K} \alpha_{2,n^{\nu}}\big(\tau(\Delta_j^{(n)}, \Delta_m^{(n)})\big) \leq$$

$$\leq C_2\, \frac{p^{3\nu}}{n^{3/2}} \Big(\frac{n}{p}\Big)^{\nu} |t|^3 f(2)\, n^{\nu\tau} \sum_{S=1}^{\infty} S^{\nu-1} \alpha(Sq) \leq$$

$$\leq C_3\, \frac{p^{2\nu} n^{\tau\nu} |t|^3 \beta(q)}{n^{1/2} q^{(2\tau+1)\nu}}.$$

The latter tends to zero as $n \to \infty$

by means of the same functions p, q guaranteeng to the tendency to zero (7.2.3) . Theorem 7.2.1 is proven.
Theorem 7.2.2 is proved analogously.

__Theorem 7.2.3.__ Let ξ_t, $t \in \mathbb{Z}^\nu$, be a centred α -mixing s.r.f. such that for some $\delta > 0$, $E|\xi_0|^{2+\delta} < \infty$, and

$$1) \quad \sum_{n=1}^{\infty} \alpha_{1,1}^{\frac{\delta}{2+\delta}} (n) < \infty ,$$

$$2) \quad \alpha_{2,\infty} (n) = 0 \left(\frac{1}{n^{2\nu}} \right) .$$

Then $\sigma^2 = \sum_{t \in \mathbb{Z}^\nu} E \xi_0 \xi_t < \infty$ and if $\sigma^2 > 0$ then for any sequence of subsets $I_n \subset \mathbb{Z}^\nu$, $n \in \mathbb{N}$, tending to infinity in a sense of Van Hove the c.l.th. holds.

__Proof.__ Let $\Gamma_s^-(I)$ be the union of parallelepipeds Λ_t such that $\Lambda_t \subset I$ and let $\Gamma_s^+(I)$ denote the union of parallelepipeds Λ_t such that $\Lambda_t \cap I \neq \varnothing$. Using conditions of Theorem 7.2.3 it is easy to check that for any $\varnothing \neq \Gamma \subset \mathbb{Z}^\nu$, $|I| < \infty$,

$$0 < C_1 |I| \leqslant DS_I \leqslant C_2 |I| . \tag{7.2.4}$$

Let I_n, $n \in \mathbb{N}$, be a sequence of subsets of $\mathbb{Z}^\nu$ tending to infinity in a sense of Van Hove. Since for any fixed $S \in \mathbb{Z}^\nu$ the relation (7.1.2) is valid then there exists a natural-valued function $\rho = \rho(n)$, $n \in \mathbb{N}$, such that $\rho \to \infty$, $\rho = 0(n)$ and $n \to \infty$ and

$$\lim_{n \to \infty} N^-(n) = +\infty, \qquad \lim_{n \to \infty} N^-(n) \left(N^+(n) \right)^{-1} = 1 .$$

Where

$$N^-(n) = N_{\hat{\rho}(n)}^-(I_n), \quad N^+(n) = N_{\hat{\rho}(n)}^+(I_n), \quad \hat{\rho}(n) = (\rho(n), \dots, \rho(n)) \in \mathbb{Z}^\nu .$$

Hence

$$\frac{|I_n| - N^-(n) \rho^\nu}{N^+(n) \rho^\nu} \leqslant \frac{N^+(n) \rho^\nu - N^-(n) \rho^\nu}{N^+(n) \rho^\nu} \to 0, \quad n \to \infty ,$$

and

$$\left(N^+(n) \rho^\nu \right)^{-1} \cdot |I_n| \to 1, \quad n \to \infty .$$

Taking into account (7.2.4) we conclude that the limiting behaviour of $\sigma_{I_n}^{-1} S_{I_n}$ is determined by $\sigma_{I_n}^{-1} S_{\Gamma_{\tilde{p}}^-(I_n)}$.

Clearly, $S_{\Gamma_{\tilde{p}}^-(I_n)}$ represents a union of ν-dimensional cubes Δ_j, $j = 1,2,\ldots, N^-(n)$ with the side of the length $p(n)$. Therefore

$$\sigma_{I_n}^{-1} S_{\Gamma_{\tilde{p}}^-(I_n)} = \sum_{j=1}^{N^-(n)} S_{\Delta_j} \sigma_{I_n}^{-1}.$$

Put

$$\hat{S}_{\Delta_j} = \frac{S_{\Delta_j^*}}{\sigma_{I_n}}, \qquad j = 1,\ldots, N^-(n),$$

where Δ_j^* denotes the cube which is concentric with Δ_j and has the side of the length $p(n) - 2q(n)$, $q(n) \to \infty$, $q = 0(p)$ as $n \to \infty$. Since according to Theorem 7.2.1 the sequence $\sigma_{\Delta_j^*}^{-1} S_{\Delta_j^*}$ is asymptotically normal then to prove our theorem it is sufficient to show that

$$\left| E \prod_{j=1}^{N^-(n)} e^{it \hat{S}_{\Delta_j}} - \prod_{j=1}^{N^-(n)} E e^{it \hat{S}_{\Delta_j}} \right| \to 0, \qquad n \to \infty,$$

which one can check in exactly the same way as those in Theorem 7.2.1.

Similarly the c.l.th. for a sequence of subsets tending to infinity in a sense of Van Hove under conditions of Theorem 7.2.2 is proved. These theorems are an improvement of some of results of Takahata /122/, Bulinskii and Zurbenko /19/ . Bolthausen's result /9/ given below also follows from Theorem 7.2.3 but we represent the original proof which uses Stein's method (Lemma 4.3.2).

<u>Theorem 7.2.4.</u> Let ξ_t, $t \in \mathbb{Z}^\nu$, be a centered s.r.f.,

$$E |\xi_0|^{2+\delta} < \infty, \quad \delta > 0 \quad \text{and}$$

$$\sum_{n=1}^\infty n^{\nu-1} \alpha_{2,\infty}^{\frac{\delta}{2+\delta}}(n) < \infty.$$

Then $\sigma^2 = \sum_{t \in \mathbb{Z}^\nu} E \xi_0 \xi_t < \infty$ and if $\sigma^2 > 0$ then for any increasing sequence of subsets $I_n \subset \mathbb{Z}^\nu$, $|I_n| < \infty$, $n \in \mathbb{N}$ such that

$$\lim_{n \to \infty} \frac{|\partial I_n|}{|I_n|} = 0$$

the c.l.th. holds.

$\underline{\text{Proof}}$. Clearly, it's sufficient to verify the conditions of Lemma 4.3.2. We have

$$\frac{1}{\sqrt{|I_n|}} \sum_{t\in I_n} |E\,\xi_t\, e^{i\lambda\left(\frac{S_{I_n}-S_{n,t}}{\sigma_0\sqrt{|I_n|}}\right)}| \leq$$

$$\leq \frac{C_1}{\sqrt{|I_n|}} |I_n|\,\alpha_{1,\infty}(p) \leq C_2\sqrt{|I_n|}\,(p(n))^{\left(\frac{2+\delta}{\delta}\right)\nu}. \tag{7.2.5}$$

Since $p(n) = 0\,(|I_n|^{\frac{1}{2\nu}})$, $n\to\infty$, the last term in (7.2.5) tends to zero as $n\to\infty$.

Further we can write

$$\frac{1}{|I_n|} \sum_{t\in I_n} E\,\xi_t\,S_{n,t}\,e^{i\lambda(\bar{S}_{I_n}-\bar{S}_{n,t})} - E\,e^{i\lambda\bar{S}_{I_n}} =$$

$$= \frac{1}{|I_n|} E\,e^{i\lambda\bar{S}_{I_n}} \sum_{t\in I_n} \xi_t\,S_{n,t}\,(e^{i\lambda\bar{S}_{n,t}}-1) +$$

$$+ E\,e^{i\lambda\bar{S}_{I_n}} \left(\frac{1}{|I_n|} \sum_{t\in I_n} \xi_t\,S_{n,t} - 1\right).$$

Using Schwarz's inequality we find

$$\frac{1}{|I_n|} \sum_{t\in I_n} |E\,\xi_t\,S_{n,t}\,(e^{-i\lambda\bar{S}_{n,t}}-1)| \leq$$

$$\leq \frac{C_1}{|I_n|^{3/2}} \sum_{t\in I_n} E\,|S_{n,t}|^2 \leq \frac{C_2\,p^\nu}{|I_n|^{1/2}} \to 0, \quad n\to\infty.$$

Denote $a_n = \sum_{t\in I_n} E\,\xi_t\,S_{n,t}$. Then

$$|E\,e^{i\lambda\bar{S}_{I_n}} \left(\frac{1}{\sigma_0|I|} \sum_{t\in I_n} \xi_t\,S_{n,t} - 1\right)| \leq$$

$$\leq C_1\,|E\,e^{i\lambda\bar{S}_{I_n}} \left(\frac{1}{a_n} \sum_{t\in I_n} \xi_t\,S_{n,t} - 1\right)| \leq \cdot$$

$$\leq a_n^{-1}\,C_1\,E^{\frac{1}{2}}\,|\sum_{t\in I_n} \xi_t\,S_{n,t} - \sum_{t\in I_n} E\,\xi_t\,S_{n,t}|^2 \leq$$

$$\leq \frac{C_1}{\delta_n}\left|\sum_{\substack{t,s,t',s'\in I_n\\ \tau(t,s)\leq p,\ \tau(t',s')\leq p}} \text{Cov}\left(\xi_t,\xi_s,\xi_{t'}\,\xi_{s'}\right)\right|^{\frac{1}{2}}.$$

If $\quad \tau(t,t') = k \geq 3p \quad$ we have

$$\left|\text{Cov}\left(\xi_t\,\xi_s,\,\xi_{t'}\,\xi_{s'}\right)\right| \leq \alpha_{2,\infty}(k-2p).$$

If min $\left(\tau(t,t'),\,\tau(t,s),\,\tau(t,s')\right) = j \quad$ then

$$\left|\text{Cov}\left(\xi_t\,\xi_s,\,\xi_{t'}\,\xi_{s'}\right)\right| \leq \left|E\,\xi_t\,\xi_s,\xi_{t'}\,\xi_{s'}\right| + \left|E\,\xi_t\,\xi_s\right|\left|E\,\xi_{t'}\,\xi_{s'}\right| \leq$$

$$\leq B\,\alpha_{1,\infty}(j)\,,\quad 0 < B < \infty.$$

Therefore

$$\frac{1}{\delta_n^2}\sum_{\substack{t,s,t',s'\in I_n\\ \tau(t,s)\leq p,\ \tau(t',s')\leq p}}\left|\text{Cov}\left(\xi_t\,\xi_s,\,\xi_{t'}\,\xi_{s'}\right)\right| \leq$$

$$\leq \frac{1}{a_n^2}|I_n|\left(p^{2\nu}\sum_{k=3p}^{\infty}k^{\nu-1}\alpha_{2,\infty}(k-2p) + C_2\,p^{2\nu}\sum_{k=0}^{3p}k^{\nu-1}\alpha_{1,\infty}(k)\right) =$$

$$= 0\left(|I_n|^{-1}p^{2\nu}\right) = 0(1)\,,\quad n \to \infty.$$

Theorem 7.2.4 is proven.

It is easy to see if $X_t^{(n)} = \left(S_t^{(n)} - E\,S_t^{(n)}\right)\left(D\,S_t^{(n)}\right)^{-\frac{1}{2}},\ t\in\mathbb{Z}^\nu,$

$n\in\mathbb{N},\ S_t^{(n)} = \sum_{j\in\Lambda_t^{(n)}}\xi_0\,,\quad \Lambda_t^{(n)} = \left\{k\in\mathbb{Z}^\nu:\,nt^{(i)}\leq k^{(i)}\leq n\,(t^{(i)}+1)\right\}$ then

under conditions of previous theorems the finite dimensional conditions of the random field $S_t^{(n)},\ t\in\mathbb{Z}^\nu,$ tends to those of $\mathcal{Z}_t,\ t\in\mathbb{Z}^\nu$, where $\mathcal{Z}_t$ are mutually independent standard normal random variables (see also /94//95/). This problem is motivated by examples from statistical physics.

Another area of results concerning the limit theorems that are uniform over a large class of sets was considered in /44/,/45/. The invariance principle for random fields with continuous parameter was studied in /50/, /77/.

7.3. Rate of Convergence in Central Limit Theorem for Random Fields

Below we will establish the convergence rate in c.l.th. which is sufficient to obtain the law of the iterated logarithm.

Theorem 7.3.1. Let ξ_t, $t \in \mathbb{Z}^\nu$, be a s.r.f., α -mixing with the coefficient $\alpha_{m,n}(\tau)$ and let

1. $E\xi_0 = 0$, $E|\xi_0|^{2+\delta} < \infty$, $\delta > 0$;

2. $\alpha_{m,n}(\tau) \leq f(m) n^\tau \alpha(\tau)$, $\tau \geq 0$,

   $f(m)$ is some non-negative function;

3. $\sum_{\tau=1}^{\infty} \tau^{\nu-1} \alpha_{1,1}^{\frac{\delta'}{2+\delta'}}(\tau) < \infty$ for some δ', $0 < \delta' < \delta$

   and

$$\alpha(\tau) = O\left(\frac{1}{\tau^{(2\tau+1)\nu+\varepsilon}}\right), \quad \varepsilon > 0.$$

Then

$$\sigma^2 = \sum_{t \in \mathbb{Z}^\nu} E\xi_0 \xi_t < \infty.$$

If in addition $\sigma^2 \neq 0$ then there exists a positive number ρ such that for the sequence of ν -dimensional cubes $I_n = [-n, n]^\nu$, $n \in \mathbb{N}$, the following estimate is valid:

$$\Delta_n = \sup_{x \in \mathbb{R}^1} \left| P(\sigma_{I_n}^{-1} S_{I_n} < x) - \Phi(x) \right| \leq \frac{C}{\ln^{1+\rho} n}, \quad n \in \mathbb{N}.$$

Proof. Let us write the following standard representation (see Section 4.1):

$$S_{I_n} = \sum_{j=1}^{K} S_{\Delta_j^{(n)}} + S_{(\bigcup_{j=1}^{K} \Delta_j^\rho)'} \tag{7.3.1}$$

$$K \sim \left[\frac{n}{\rho}\right]^\nu, \quad n \to \infty.$$

Then we have

$$\left| E e^{it\sigma_{I_n}^{-1} S_{I_n}} - e^{-\frac{t^2}{2}}\right| \le \frac{|t|}{\sigma_{I_n}} E\left| S_{\left(\bigcup_{j=1}^{K} \Delta_j^{(n)}\right)'}\right| + \tag{7.3.2}$$

$$+ \left| E e^{it\sigma_{I_n}^{-1}\sum_{j=1}^{K} S_{\Delta_j^{(n)}}} - e^{-\frac{t^2}{2}}\right| \le C_l |t| \sqrt{\frac{q}{p}} +$$

$$+ \left| E e^{it\sum_{j=1}^{K} S(j,n)} - e^{-\frac{t^2}{2}}\right|, \qquad S(j,n) = \frac{S_{\Delta_j^{(n)}}}{\sigma_{I_n}}, \qquad j = \overline{1,k}.$$

Let's introduce the following notations:

$$f_N(x) = \begin{cases} x, & |x| \le N \\ 0, & |x| > N \end{cases}, \qquad \overline{f}_N(x) = x - f_N(x),$$

$$S_{I_n}^{(N)} = \sum_{t \in I_n} f_N(\xi_t), \qquad \overline{S}_{I_n}^{(N)} = \sum_{t \in I_n} \overline{f}_N(\xi_t),$$

$$S_{\Delta_j^{(n)}}^{(N)} = \sum_{t \in \Delta_j^{(n)}} f_N(\xi_t), \qquad \overline{S}_{\Delta_j^{(n)}}^{(N)} = \sum_{t \in \Delta_j^{(n)}} \overline{f}_N(\xi_t),$$

$$S^{(N)}(j,n) = \sigma_{I_n}^{-1} S_{\Delta_j^{(n)}}^{(N)}, \qquad \overline{S}^{(N)}(j,n) = \sigma_{I_n}^{-1} \overline{S}_{\Delta_j^{(n)}}^{(N)}$$

$$\sigma_0^2(N) = E f_N(\xi_0)^2 - (E f_N(\xi_0))^2 + \sum_{t \in \mathbb{Z}^\nu \backslash \{0\}} \left(E f_N(\xi_0) f_N(\xi_t) - E f_N(\xi_0) E f_N(\xi_t)\right),$$

$$\sigma_{I_n}(N) = D S_{I_n}^{(N)}, \qquad \overset{*}{S}^{(N)}(j,n) = \sigma_{I_n}^{-1}(N) S_{\Delta_j^{(n)}}^{(N)}.$$

Further

$$\left| E e^{it\sum_{j=1}^{K} S(j,n)} - e^{-\frac{t^2}{2}}\right| = \left| E e^{it\sum_{j=1}^{K} \left(S^{(N)}(j,n) + \overline{S}^{(N)}(j,n)\right)} - e^{-\frac{t^2}{2}}\right| \le$$

$$\le |t| E \left| \sum_{j=1}^{K} \overline{S}^{(N)}(j,n)\right| + \left| E e^{it\sum_{j=1}^{K} S^{(N)}(j,n)} - e^{-\frac{t^2}{2}}\right| \le$$

$$\leq |t|\, E\,|\sum_{j=1}^{K} \bar{S}^{(N)}(j,n)| + | E\, e^{\,it\sum_{j=1}^{K} S^{(N)}(j,n)} -$$

$$- E\, e^{\,it\sum_{j=1}^{K} \overset{*}{S}^{(N)}(j,n)} | + | E\, e^{\,it\sum_{j=1}^{K} \overset{*}{S}^{(N)}(j,n)} - e^{-\frac{t^2}{2}} | \leq$$

$$\leq |t|\, E\,|\sum_{j=1}^{K} \bar{S}^{(N)}(j,n)| + |t|\,|\frac{\sigma_{I_n}(N)}{\sigma_{I_n}} - 1| +$$

$$+ | E\, e^{\,it\sum_{j=1}^{K} \overset{*}{S}^{(N)}(j,n)} - \prod_{j=1}^{K} E\, e^{\,it\overset{*}{S}^{(N)}(j,n)} | +$$

$$+ | \prod_{j=1}^{K} E\, e^{\,it\overset{*}{S}^{(N)}(j,n)} - e^{-\frac{t^2}{2}} | .$$

In just the same way as in the proof of Theorem 7.2.3 one may obtain the following estimate

$$| E\, e^{\,it\sum_{j=1}^{K} \overset{*}{S}^{(N)}(j,n)} - \prod_{j=1}^{K} E\, e^{\,it\overset{*}{S}^{(N)}(j,n)} | \leq \qquad\qquad (7.3.3.)$$

$$\leq C_2 \Big(N|t|^3 \big(\frac{p}{n^{\frac{1}{2}}}\big)^{\nu} + (Nt)^2 \frac{p^{\nu} n^{\tau\nu}}{q^{(2\tau+1)\nu+\varepsilon}} + (Nt)^3 \frac{p^{2\nu} n^{\tau\nu}}{n^{\nu/2} q^{(2\tau+1)\nu+\varepsilon}} \Big) .$$

Using Theorem 1.6.5 we have

$$| \prod_{j=1}^{K} E\, e^{\,it\overset{*}{S}^{(N)}(j,n)} - e^{-\frac{t^2}{2}} | \leq$$

$$\leq 16|t|^3 e^{-\frac{t^2}{2}} \Big(\sum_{j=1}^{K} \frac{DS^{(N)}\Delta_j^{(n)}}{\sigma_{I_n}^2(N)} \Big)^{-\frac{3}{2}} L_n^{(N)} , \qquad |t| \leq \frac{1}{4 L_n^{(N)}} ,$$

$$L_n^{(N)} = \sum_{j=1}^{K} E\,|\frac{S^{(N)}\Delta_j^{(n)}}{\sigma_{I_n}(N)}|^3 \leq \frac{C_3 N n^{\nu} p^{\nu}}{\sigma_{I_n}^3(N)} .$$

Since

$$\sigma_{I_n}^2(N) \geq \frac{\sigma^2(N)}{2} n^{\nu} (1 + 0(1)) , \qquad n \to \infty ,$$

then for $|t| \leq \dfrac{1}{4 L_n^{(N)}}$

$$\left| \prod_{j=1}^{K} E\, e^{it \overset{*}{S}{}^{(N)}(j,n)} - e^{-\frac{t^2}{2}} \right| \leq C_4 |t|^3 e^{-\frac{t^2}{3}} N \left(\frac{P}{n^{1/2}} \right)^{\nu}. \tag{7.3.4}$$

Applying Theorem 3.1.9 we find (below $\varepsilon = \min\{\frac{1}{2}, \frac{(\delta-\delta')}{\delta'(2+\delta)}\}$)

$$\left| \sigma_{I_n}^{-1} \sigma_{I_n}(N) - 1 \right| = \left| \frac{\sigma_{I_n}^2(N) - \sigma_{I_n}^2}{\sigma_{I_n}(\sigma_{I_n}(N) + \sigma_{I_n})} \right| =$$

$$= \left| \frac{\sigma_0^2 n^{\nu} - \sigma_0^2(N) n^{\nu} + O(n^{\nu - \varepsilon})}{\sigma_{I_n}(\sigma_{I_n}(N) + \sigma_{I_n})} \right| \leq$$

$$\leq C_5 \left| \sigma_0^2 - \sigma_0^2(N) \right| \leq C_5 \left(E \left| f_N^2(\xi_0) - \xi_0^2 \right| + \left(E f_N(\xi_0) \right)^2 + \right.$$

$$\left. + \sum_{t \in \mathbb{Z}^{\nu} \setminus \{0\}} \left| E \xi_0 \xi_t - E f_N(\xi_0) f_N(\xi_t) + E f_N(\xi_0) E f_N(\xi_j) \right| \right) \leq$$

$$\leq C_5 \left(E \left| f_N(\xi_0) - \xi_0 \right| \left| f_N(\xi_0) + \xi_0 \right| + \left(E |\xi_0| I(|\xi_0| > N) \right)^2 + \right.$$

$$\left. + \sum_{t \in \mathbb{Z}^{\nu} \setminus \{0\}} \left| E f_N(\xi_0) \bar{f}_N(\xi_t) + E \bar{f}_N(\xi_0) f_N(\xi_t) + \right. \right.$$

$$\left. \left. + E \bar{f}_N(\xi_0) \bar{f}_N(\xi_t) + E f_N(\xi_0) E f_N(\xi_t) \right| \right) \leq$$

$$\leq C_6 \left(E^{\frac{1}{2}} |\xi_0|^2 I(|\xi_0| \geq N) + \frac{1}{N^{2(1+\delta)}} + \sum_{t \in \mathbb{Z}^{\nu} \setminus \{0\}} \left(E \bar{f}_N(\xi_0) f_N(\xi_t) - \right. \right.$$

$$\left. - E \bar{f}_N(\xi_0) E f_N(\xi_t) + E f_N(\xi_0) \bar{f}_N(\xi_t) - E f_N(\xi_0) E \bar{f}_N(\xi_t) + \right.$$

$$\cdot + E\,\bar{f}_N(\xi_0)\,\bar{f}_N(\xi_t) - E\,\bar{f}_N(\xi_0)\,E\,\bar{f}_N(\xi_t)) \leqslant$$

$$\leqslant C_7\Big(\frac{1}{N^{\delta/2}} + E^{\frac{1}{2+\delta'}}\,|\xi_0|^{2+\delta'}\,I(|\xi_0|\geqslant N)\Big) \leqslant$$

$$\leqslant C_8\Big(\frac{1}{N^{\delta/2}} + \frac{1}{N^{\frac{\delta-\delta'}{2+\delta'}}}\Big) \leqslant C_9\,N^{-\frac{\delta-\delta'}{2+\delta'}}.$$

Note also that

$$E\Big|\sum_{j=1}^{K}\bar{S}^{(N)}(j,n)\Big|^2 \leqslant C_{10}\Big(\frac{A\,E|\bar{f}_N(\xi_0)|^2}{\sigma^2} + \sum_{j=A+1}^{\infty} j^{\nu-1}\alpha^{\frac{\delta}{2+\delta}}(j)\Big) \leqslant$$

$$\leqslant C_{11}\Big(\frac{A}{N^{\delta}} + \sum_{j=A+1}^{\infty} j^{\nu-1} j^{-\frac{\delta(2+\delta')\nu}{(2+\delta)\delta'}}\Big) \leqslant$$

$$\leqslant C_{12}\Big(\frac{A}{N^{\delta}} + N^{\frac{2(\delta'-\delta)}{\delta'(2+\delta)}}\Big) \leqslant C_{13}\,N^{\frac{2(\delta'-\delta)}{\delta'(2+\delta)}}, \qquad A\in\mathbb{N}.$$

Hence

$$E\Big|\sum_{j=1}^{K}\bar{S}^{(N)}(j,n)\Big|^2 \leqslant C_{13}\,N^{\frac{2(\delta'-\delta)}{\delta'(2+\delta)}}. \tag{7.3.5}$$

Finally using (7.3.2)–(7.3.5) we obtain

$$\Big|E\,e^{it\sigma_n^{-1}S_{In}} - e^{-\frac{t^2}{2}}\Big| \leqslant C_{14}\Big(|t|\sqrt{\frac{q}{p}} + |t|\,N^{-\frac{\delta-\delta'}{2+\delta'}} +$$

$$+ N|t|^3\Big(\frac{p}{n^{1/2}}\Big)^{\nu} + \frac{N^2 t^2 n^{\tau\nu} p^{\nu}}{q^{(2\tau+1)\nu+\varepsilon}} + N^3 t^3\frac{p^{2\nu} n^{\tau\nu}}{n^{\nu/2} q^{(2\tau+1)+\varepsilon}} +$$

$$+ |t|^3 e^{-\frac{t^2}{2}} N\Big(\frac{p}{n^{1/2}}\Big)^{\nu}, \qquad |t| \leqslant \Big(C_{15}\,N\Big(\frac{p}{n^{1/2}}\Big)^{\nu}\Big)^{-1}.$$

Putting

$$T = (\ln n)^{1+\rho}, \qquad N = (\ln n)^{2(1+\rho)\frac{2+\delta}{\delta-\delta'}},$$

$$p(n) = \left[n^{\frac{1}{2}} (\ln n)^{-\frac{2(1+\rho)(2+\frac{2+\delta}{\delta-\delta'})}{\nu}} \right],$$

$$q(n) = \left[n^{\frac{1}{2}} (\ln n)^{-\frac{\nu 4(1+\rho)+2(1+\rho)(2+\frac{2+\delta}{\delta-\delta'})}{\nu}} \right], \quad \rho > 0$$

with the help of Berry-Esseen inequality we obtain

$$\Delta_n \leqslant C \left[T\sqrt{\frac{q}{p}} + TN^{\frac{\delta'-\delta}{2+\delta}} + NT^3 \left(\frac{p}{n^{1/2}}\right)^\nu + N\left(\frac{p}{n^{1/2}}\right)^\nu + \right.$$

$$\left. + \frac{N^2 T^2 p^\nu n^{\tau\nu}}{q^{(2\tau+1)\nu+\varepsilon}} + \frac{N^3 T^3 p^{2\nu} n^{\tau\nu}}{n^{\frac{1}{2}} q^{(2\tau+1)\nu+\varepsilon}} + \frac{1}{T} \right] \leqslant \frac{C}{\ln^{1+\rho} n}, \quad \rho > 0.$$

Theorem 7.3.1 is proven.

Theorem 7.3.2. Let ξ_t, $t \in \mathbb{Z}^\nu$, be a φ-mixing random field with coefficient $\varphi_{n,\infty}(\tau)$ and let

1. $E\xi_0 = 0, \qquad E|\xi_0|^{2+\delta} < \infty, \quad \delta > 0$,

2. $\varphi_{n,\infty}|\tau| \leqslant f(n)\varphi(\tau), \qquad \sum_{\tau=1}^{\infty} \tau^{\nu-1}\varphi^{\frac{1}{2}}(\tau) < \infty.$

Then

$$\sigma^2 = \sum_{t \in \mathbb{Z}^\nu} E\,\xi_0 \xi_t < \infty.$$

If in addition $\sigma^2 \neq 0$ then there exists a positive constant ρ such that for the sequence of ν-dimensional cubes $I_n = [-n, n]^\nu$, $n \in \mathbb{N}$, the next estimate is valid:

$$\Delta_n = \sup_{x \in R^1} |P(\sigma_{I_n}^{-1} S_{I_n} < x) - \Phi(x)| \leqslant \frac{C}{\ln^{1+\rho} n}, \quad \rho > 0.$$

168

<u>Proof.</u>   Using representation (7.3.1) and inequality (1.6.1) we can write

$$P\Big(\sum_{j=1}^{K}\hat{S}_{\Delta_{j}^{(n)}}<x-\varepsilon\Big)-\frac{1}{\varepsilon^{2}}\,\sigma_{I_{n}}^{-2}\,DS_{(\bigcup_{j=1}^{K}\Delta_{j}^{(n)})'}\le$$

$$\le P\big(\sigma_{I_{n}}^{-1}S_{I_{n}}<x\big)\le P\Big(\sum_{j=1}^{K}\hat{S}_{\Delta_{j}^{(n)}}<x+\varepsilon\Big)+\frac{1}{\varepsilon^{2}}\sigma_{I_{n}}^{-2}\,DS_{(\bigcup_{j=1}^{K}\Delta_{j}^{(n)})'}\,,$$

$$\hat{S}_{\Delta_{j}^{(n)}}=\sigma_{I_{n}}^{-1}\,S_{\Delta_{j}^{(n)}}$$

Hence

$$P\Big(\sum_{j=1}^{K}\hat{S}_{\Delta_{j}^{(n)}}<x-\varepsilon\Big)-\frac{C_{1}q}{\varepsilon^{2}p}\le P\big(\sigma_{I_{n}}^{-1}S_{I_{n}}<x\big)\le$$

$$\le P\Big(\sum_{j=1}^{K}S_{\Delta_{j}^{(n)}}<x+\varepsilon\Big)+\frac{C_{1}q}{\varepsilon^{2}p}\,.$$

According to inequality (7.2.1) we have

$$\Big|E\prod_{j=1}^{K}e^{it\hat{S}_{\Delta_{j}^{(n)}}}-\prod_{j=1}^{K}Ee^{it\hat{S}_{\Delta_{j}^{(n)}}}\Big|\le T_{1}+T_{2}+T_{3}\,.$$

Further

$$T_{1}\le 2\,\frac{|t|^{2+\delta}}{3!}\sum_{j=1}^{K-1}E|\hat{S}_{\Delta_{j}^{(n)}}|^{2+\delta}\le C_{2}|t|^{2+\delta}\frac{p^{\nu(2+\delta)}}{n^{\frac{\nu(2+\delta)}{2}}}\Big(\frac{n}{p}\Big)^{\nu}=$$

$$=C_{2}|t|^{2+\delta}p^{\nu(1+\delta)}n^{-\frac{\nu\delta}{2}}\,.$$

Applying Point 2) of Lemma 2.1.1   for $\tau=2+\delta$, $s=2$
$\varepsilon=\delta\big(2(2+\delta)\big)^{-1}$ **we find**

$$T_{2}\le|t|\,\sigma_{I_{n}}^{-1}\sum_{j=1}^{K-1}\sum_{s\in\Delta_{j}^{(n)}}\sum_{m=j+1}^{K}\Big|E\,\xi_{s}\big(e^{it\hat{S}_{\Delta_{m}^{(n)}}}-1\big)\prod_{\tau=m+1}^{K}e^{it\hat{S}_{\Delta_{\tau}^{(n)}}}-$$

$$-E\xi_s E(e^{it\hat{S}_{\Delta_m^{(n)}}}-1)\prod_{\tau=m+1}^{K}e^{it\hat{S}_{\Delta_\tau^{(n)}}}|\leq$$

$$\leq|t|^2\sigma_{I_n}^{-1}\sum_{j=1}^{K-1}\sum_{s\in\Delta_j^{(n)}}\sum_{m=j+1}^{K}E^{\frac{1}{2+\delta}}|\xi_s|^{2+\delta}E^{\frac{1}{2}}|\hat{S}_{\Delta_m^{(n)}}|^2\varphi^{\frac{1+\delta}{2+\delta}}(\rho(\Delta_j^{(n)},\Delta_m^{(n)}))\leq$$

$$\leq C_3|t|^2 f(1)\frac{p^2\cdot p^{\frac{\nu}{2}}}{\sigma_{I_n}^2}\sum_{j=1}^{K-1}\sum_{m=j+1}^{K}\varphi^{\frac{1+\delta}{2+\delta}}(\rho(\Delta_j^{(n)},\Delta_m^{(n)}))\leq$$

$$\leq\frac{C_4|t|^2 f(1)p^{\frac{3\nu}{2}}}{n^\nu}\left(\frac{n}{p}\right)^\nu\sum_{j=1}^{K}\varphi^{\frac{1+\delta}{2+\delta}}(\rho(\Delta_j^{(n)},\Delta_m^{(n)}))\leq$$

$$\leq C_5|t|^2 f(1)p^{\frac{\nu}{2}}\sum_{j=1}^{\infty}j^{\nu-1}\varphi^{\frac{1+\delta}{2+\delta}}(jq)\leq$$

$$\leq C_6|t|^2 f(1)\frac{p^{\frac{\nu}{2}}}{q^{\frac{2\nu(1+\delta)}{2+\delta}}}\sum_{j=1}^{\infty}\frac{1}{j^{1+\frac{\nu\delta}{2+\delta}}}\leq C_7|t|^2 p^{\frac{\nu}{2}}q^{-\frac{2\nu(1+\delta)}{2+\delta}}.$$

Further by Corollary 2.1.1 we have

$$T_3\leq\sigma_{I_n}^{-2}\frac{t^2}{2}\sum_{j=1}^{K-1}\sum_{m=j+1}^{K}\sum_{t,e\in\Delta_j^{(n)}}|E\xi_t\xi_e(e^{it\hat{S}_{\Delta_{j+1}^{(n)}}}-1)\prod_{m=j+2}^{K}e^{it\hat{S}_{\Delta_m^{(n)}}}-$$

$$-E\xi_t\xi_e E(e^{it\hat{S}_{\Delta_{j+1}^{(n)}}}-1)\prod_{m=j+2}^{K}e^{it\hat{S}_{\Delta_m^{(n)}}}|\leq$$

$$\leq C_8\sigma_{I_n}^{-2}\frac{t^2 p^{2\nu}}{2}\sum_{j=1}^{K-1}\sum_{m=j+1}^{K}\varphi(\rho(\Delta_j^{(n)},\Delta_m^{(n)}))\leq$$

$$\leq C_9\frac{t^2 p^{2\nu}}{n^\nu}\left(\frac{n}{p}\right)^\nu\sum_{j=1}^{K}\varphi(\rho(\Delta_0^{(n)},\Delta_j^{(n)}))\leq$$

$$\leq C_9\frac{t^2 p^{2\nu}}{n^\nu}\left(\frac{n}{p}\right)^\nu\sum_{j=1}^{K}j^{\nu-1}\varphi(jq)\leq$$

$$\leq C_9 \frac{t^2 p^\nu}{q^{2\nu}} \sum_{j=1}^{\infty} \frac{1}{j^{1+\nu}} \leq C_{10}\, t^2 p^\nu q^{-2\nu}.$$

Finally

$$\left| E \prod_{j=1}^{K} e^{it\hat{S}_{\Delta_j^{(n)}}} - \prod_{j=1}^{K} E\, e^{it\hat{S}_{\Delta_j^{(n)}}} \right| \leq C_{11}\left(|t|^{2+\delta}\, p^{\nu(1+\delta)}\, n^{-\frac{\nu\delta}{2}} + \right.$$

$$\left. + |t|^2 p^{\frac{\nu}{2}} q^{-\frac{2\nu(1+\delta)}{2+\delta}} + |t|^2 p^\nu q^{-2\nu} \right). \qquad (7.3.6)$$

Now let Y_j, $j = \overline{1,n}$, be a sequence of independent random variables each of which is distributed as $\hat{S}_{\Delta_0^{(n)}}$. Hence using (7.3.6) and the Berry-Esseen inequality we can write

$$\Delta^n(x) = \left| P\left(\sum_{j=1}^{K} \hat{S}_{\Delta_j^{(n)}} < x + \varepsilon \right) - P\left(\sum_{j=1}^{K} Y_j < x + \varepsilon \right) \right| \leq$$

$$\leq C_{12}\left(T^{2+\delta}\, p^{\nu(1+\delta)}\, n^{-\frac{\nu\delta}{2}} + T^2 p^{\frac{\nu}{2}} q^{-\frac{2\nu(1+\delta)}{2+\delta}} + T^2 p^\nu q^{-2\nu} + \frac{1}{T} \right).$$

Putting

$$T = \ell n^{1+\rho} n, \quad p = n^\gamma, \quad q = n^\beta, \quad 0 < \beta < \gamma < \frac{\delta}{2(1+\delta)},$$

we obtain

$$\Delta^{(n)}(x) \leq \frac{C_{13}}{\ell n^{1+\rho} n}, \quad \rho > 0.$$

Hence we have

$$P\left(\delta_{I_n}^{-1} S_{I_n} < x \right) - \Phi(x) \leq \frac{C_{13}}{\ell n^{1+\rho} n} + \left| P\left(\sum_{j=1}^{K} Y_j < x + \varepsilon \right) - \right.$$

$$\left. - \Phi(x+\varepsilon) \right| + \left| \Phi(x+\varepsilon) - \Phi(x) \right| + \frac{C_1 q}{\varepsilon^2 p} \leq$$

$$\leqslant \frac{C_{13}}{\ell n^{1+\rho} n} + \frac{C_{14}}{n^{\nu(\frac{2+\delta}{2})}} \sum_{j=1}^{K} E|S_{\Delta_j^{(n)}}|^{2+\delta} + C_{15}\,\varepsilon.$$

Here we applied Theorem 1.6.2 . Putting $\varepsilon = \dfrac{1}{\ell n^{1+\rho} n}$ we find

$$P(\sigma_{I_n}^{-1} S_{I_n} < x) \leqslant C_{16} \left(\frac{1}{\ell n^{1+\rho} n} + \frac{\rho^{\nu(1+\delta)}}{n^{\frac{\nu\delta}{2}}} \right) \leqslant \frac{C}{\ell n^{1+\rho} n}.$$

In exactly the same way one can obtain the required lower bound.

Note, recently Sunklodas /121/ has obtained various estimates of the convergence rate for the α- and φ -mixing random fields such that

$$\alpha_{m,n}(\imath) \leqslant (m+n)^\rho \alpha(\imath), \qquad \rho > 0 \ ; \quad m,n \in \mathbb{N}$$

or

$$\varphi_{m,n}(\imath) \leqslant (m+n)^\rho \varphi(\imath), \qquad \rho > 0 \ ; \quad m,n \in \mathbb{N}$$

obtained various estimates of the convergence rate in c.l.th. Below we formulate some of these results.

<u>Theorem 7.3.3.</u> Suppose a centered s.r.f. ξ_t, $t \in \mathbb{Z}^\nu$, satisfies α -mixing condition, $E|\xi_0|^s < \infty$, $2 < s \leqslant 3$ and

$$\alpha(\imath) \leqslant K\,\imath^{-\mu},$$

where

$$\mu = \frac{2\nu(s-1)}{s(s-2)^2}\left[B(5s-6)-(s-2)(3-s)+\rho(s-2)(2B+s-2)\right],$$
$$0 < K < \infty, \quad B > 1.$$

Assume also that

$$DS_V \geqslant C|V|, \quad C > 0, \quad V \subset \mathbb{Z}^\nu.$$

Then for any ρ such that $0 \leqslant \rho \leqslant \dfrac{2s}{s-2}$

$$\Delta_V = \sup_{x \in R^1} |P(\frac{S_V}{\sqrt{DS_V}} < x) - \Phi(x)| \leqslant$$

$$\leqslant C(K, B, \nu, \rho, s)\,E|\xi_0|^s\,(DS_V)^{-\frac{(B-1)(s-2)}{2B+s-2}}.$$

<u>Theorem 7.3.4.</u> Suppose a centered s.r.f. ξ_t, $t \in \mathbb{Z}^\nu$, satisfies φ -mixing condition, $E|\xi_0|^s < \infty$, $2 < s \leqslant 3$ and

$$\varphi(\tau) \leqslant K \tau^{-\mu},$$

where

$$\mu = \frac{2\nu(s-1)}{s^2(s-2)}\left(4B(2s-3) - (s-2)(6-s) + ps(2B+s-2)\right),$$

$$0 < K < \infty, \quad B > 1.$$

Assume also that $DS_V \geqslant C|V|$, $C > 0$.

Then for any p, $0 \leqslant p < \infty$

$$\int_{-\infty}^{\infty} |P\left(\frac{S_V}{\sqrt{DS_V}} < x\right) - \Phi(x)|\, dx \leqslant C(K, B, \nu, p, s)\, E|\xi_0|^s (DS_V)^{-\frac{(B-1)(s-2)}{2B+s-2}}.$$

See also a very general results of Bulinskii /18/ and /135/.

7.4 Law of Iterated Logarithm for Random Fields

Here will be considered the random fields ξ_t, $t \in \mathbb{Z}^\nu$, such that $DS_{[-n,n]^\nu} = \sigma^2 (2n)^\nu (1 + 0\,(1))$. We say that a s.r.f. $\xi_t, t \in \mathbb{Z}^\nu$, satisfies the law of the iterated logarithm if the following relation is valid

$$P\left(\lim_{n \to \infty} \sup \frac{S_{[-n,n]^\nu}}{\sigma^2 \sqrt{2(2n)^\nu \log\log (2n)^\nu}} = 1\right) = 1.$$

The next theorem establishes some general conditions under which for a random field the law of the iterated logarithm holds.

<u>Theorem 7.4.1.</u> Let ξ_t, $t \in \mathbb{Z}^\nu$, be a centered, α -mixing s.r.f. such that

1. $\alpha_{m,n}(\tau) \leqslant m^{\tau_1} n^{\tau_2} \alpha(\tau)$, $\quad \alpha(\tau) = O\left(\frac{1}{\tau^{\beta\nu}}\right)$, $\quad \beta > \tau_1 + \tau_2$, $\quad \tau_1 \geqslant 0, \quad \tau_2 \geqslant 0$

2) $\sigma_n^2 = DS_{I_n} = \sigma^2 (2n)^\nu (1 + 0(1))$, $\quad n \to \infty$, $\quad I_n = [-n, n]^\nu$,

3) $\sup_{x \in R^1} |P(\sigma_n^{-1} S_{I_n} < x) - \Phi(x)| = O\left(\frac{1}{\log^{1+p} n}\right)$, $\quad p > 0$,

173

4) $P(\max_{1\le j\le n}|S_{I_j}| \ge \beta\chi(n)) = O(\frac{1}{\log^{1+\rho}n})$, $\rho>0$,

$$\chi(n) = ((2n)^\nu \log\log(2n)^\nu)^{1/2}, \qquad \beta>1.$$

Then for the random field ξ_t, $t\in\mathbb{Z}^\nu$, the law of the iterated logarithm holds.

Proof. As in one-dimensional case it is sufficient to show that for any $\varepsilon>0$

$$P(|S_{I_n}| > (1+\varepsilon)\chi(n), \quad i.o.) = 0, \tag{7.4.1}$$

$$P(|S_{I_n}| > (1-\varepsilon)\chi(n), \quad i.o.) = 1. \tag{7.4.2}$$

The relation (7.4.1) may be proved in just the same way as it was made for the random process. Let us prove (7.4.2). Put

$$E_K = \{|S_{p^i}| \le (1-\delta)\chi(p^i), \quad i=1,2,\ldots, K-1\};$$

$$F_K = \{|S_{p^K}| > (1-\delta)\chi(p^K)\};$$

$$G_K = \{|S_{I_{p^K}} - S_{I_{p^{K-1}+n_K}}| > (1-\gamma)\chi(p^K)\};$$

$$L_K = \{|S_{I_{p^K+n_K}} - S_{I_{p^{K-1}}}| < \varepsilon\chi(p^K)\}.$$

Here $K\in\mathbb{N}$, $n_K=[p^{\frac{K}{2}}]$ and γ is such that

$$\frac{2}{\sqrt{p}} + \gamma + \varepsilon < \delta.$$

Denote

$$u_K = P(E_K\cap F_K), \qquad \mathcal{U}_m = \sum_{K=1}^m u_K.$$

By Condition 3) we have

$$P(|S_{I_n}| > \beta\chi(n)) \ge \frac{c_1}{(\log n)(\log\log n)}$$

174

and since for all sufficiently large

$$m_\kappa = p^\kappa - (p^{\kappa-1} + m_\kappa) > \left[\frac{p^\kappa}{2}\right],$$

then we can write

$$v_\kappa = P(G_\kappa) \geq P(|S_{I_{m_\kappa}}| > 2(1-\gamma)\, \chi\,([\tfrac{p^\kappa}{2}])) \geq$$

$$\geq P(|S_{I_{m_\kappa}}| > 2(1-\gamma)\, \chi\,(m_\kappa)) > \frac{C_2}{k\,\log k}\,.$$

Using Chebyshev's inequality we have

$$P(L_\kappa) \geq 1 - C_3\, p^{-\frac{\kappa\nu}{2}}\,.$$

Further

$$P(E_\kappa \cap G_\kappa) = u_\kappa = \mathcal{U}_\kappa - \mathcal{U}_{\kappa-1} \geq P(E_\kappa \cap G_\kappa \cap L_\kappa) \geq$$

$$\geq P(E_\kappa \cap G_\kappa) - P(\bar{L}_\kappa) \geq P(E_\kappa)P(G_\kappa) - \alpha_{I_{p^\kappa} I_{p^{\kappa-1}}}(p^{\frac{\kappa}{2}}) - P(\bar{L}_\kappa).$$

Since

$$P(E_\kappa) \geq 1 - \mathcal{U}_\kappa$$

then

$$\mathcal{U}_\kappa \geq \mathcal{U}_{\kappa-1} + v_\kappa(1 - \mathcal{U}_{\kappa-1}) - \alpha_{I_{p^\kappa} I_{p^{\kappa-1}+n_\kappa}}(p^{\frac{\kappa}{2}}) - C_3\, p^{-\frac{\kappa\nu}{2}}$$

and

$$1 - \mathcal{U}_\kappa \leq (1 - v_\kappa)(1 - \mathcal{U}_{\kappa-1}) + \alpha_{p^{\nu\kappa},\,(p^{\kappa-1}+n_\kappa)^\nu}(p^{\frac{\kappa}{2}}) + C_3\, p^{-\frac{\kappa\nu}{2}}\,.$$

Finally

$$\sum_{\kappa \geq 1} p^{\nu\kappa\tau_1}(p^{\kappa-1}+n_\kappa)^{\nu\tau_2}\frac{1}{p^{\nu\beta\kappa}} + C_3 \sum_{\kappa \geq 1} p^{-\frac{\kappa\nu}{2}} \to 0,\quad n \to \infty\,.$$

Theorem 7.4.1 is proven.

$\underline{\text{Theorem 7.4.2.}}$ Let ξ_t, $t\in\mathbb{Z}^\nu$, be an α-mixing s.r.f. with mixing coefficient $\alpha_{m,n}(\tau)$ and let

$$1. \quad E\xi_0 = 0, \quad E|\xi_0|^{2+\delta} < \infty, \quad \delta > 0;$$

$$2. \quad \alpha_{m,n}(\tau) \leq m^{\tau_1} n^{\tau_2} \alpha(\tau), \quad \tau_1, \tau_2 \geq 0;$$

3. For some δ', $0 < \delta' < \delta$ the series

$$\sum_{\tau=1}^{\infty} \tau^{\nu-1} \alpha^{\frac{\delta'}{2+\delta'}}(\tau)$$

converges and $\alpha(\tau) = O\left(\frac{1}{\tau^{\beta\nu}}\right)$, $\beta > \frac{\nu}{\nu-1}(\tau_1+\tau_2) + \frac{1}{\nu(\nu-1)}$

Then $\sigma^2 = \sum_{t\in\mathbb{Z}^\nu} E\xi_0\xi_t < \infty.$

If in addition $\sigma^2 \neq 0$ then for the random field ξ_t, $t\in\mathbb{Z}^\nu$, the law of the iterated logarithm holds.

$\underline{\text{Proof.}}$ It is sufficient to check Condition 4 of Theorem 7.4.1. Putting $x_j = S_{I_j \setminus I_{j-1}}$, $j = \overline{1,n}$, $S_{I_0} = 0$ and using inequality (3.3.1) we have

$$P\left(\max_{1\leq i\leq n} |S_{I_i}| \geq \beta\chi(n)\right) \leq \left(P(|S_{I_n}| \geq \frac{\beta\chi(n)}{3}) + \right.$$

$$\left. + \alpha(p)\sum_{j=1}^{k-2}\left((i+2)p\right)^{\nu\tau_1}\left(n^\nu - ((i+2)p)^\nu\right)^{\tau_2} + \frac{n}{p}P\left(\sum_{j=1}^{2p}|S_{I_j} - S_{I_i}| \geq \frac{\beta\chi(n)}{3}\right)\right) \times$$

$$\times \left(1 - \max_{1\leq i\leq k-2} P(|S_{I_n} - S_{I_{(i+2)p}}| \geq \frac{\beta\chi(n)}{3})\right)^{-1}.$$

Further

$$P(|S_{I_n}| \geq \frac{\beta\chi(n)}{3}) \leq |P(\sigma_{I_n}^{-1} S_{I_n} \geq \frac{\beta\chi(n)}{3\sigma_{I_n}}) + $$

$$+ \Phi\left(\frac{\beta\chi(n)}{3\sigma_{I_n}}\right) - 1| + |P(\sigma_{I_n}^{-1} S_{I_n} \leq -\beta\frac{\chi(n)}{3\sigma_{I_n}}) - $$

$$- \Phi\left(-\frac{\beta\chi(n)}{3\sigma_{I_n}}\right)| + 2\Phi\left(-\beta\frac{\chi(n)}{3\sigma_{I_n}}\right) \leq$$

176

$$\leq C_1 \, \Delta_n + C_2 \, e^{-\beta_1^2 \log\log n} \leq$$

$$\leq \frac{C_3}{\log^{1+\rho} n} + \frac{C_4}{(\log n)^{\beta_1^2}} \, , \qquad \beta_1 > 1 \, .$$

Putting $\quad p = \left[\dfrac{n^{\frac{\nu-1}{\nu}}}{\log^{\frac{1+\rho}{\nu}} n} \right] \quad$ we can write

$$\alpha(p) \sum_{j=1}^{k-2} \left((i+1)p \right)^{\nu \tau_1} \left(n^\nu - \left((i+2)p \right)^\nu \right)^{\tau_2} \leq$$

$$\leq \alpha(p) \, p^{\nu \tau_1} \, n^{\nu \tau_2} \sum_{j=1}^{k-2} (i+1)^{\nu \tau_1} \leq$$

$$\leq C_5 \, \alpha(p) \, p^{\nu \tau_1} \, n^{\nu \tau_2} \left(\frac{n}{p} \right)^{\nu \tau_1 + 1} \leq$$

$$\leq C_6 \, \frac{n^{\nu(\tau_1 + \tau_2) + 1}}{p^{\beta \nu + 1}} \leq \frac{C_7}{\log^{1+\rho} n} \, .$$

Further

$$\frac{n}{p} P\left(\sum_{j=1}^{2p} |S_{I_j} - S_{I_{j-1}}| \geq \frac{\beta \chi(n)}{3} \right) \leq \frac{C_8 \, D\left(\sum_{j=1}^{2p} |S_{I_j} - S_{I_{j-1}}| \right)}{n^\nu \log\log n^\nu} \cdot \frac{n}{p}$$

$$\leq \frac{C_8 \left(\sum_{i=1}^{2p} \sqrt{D(S_{I_j} - S_{I_{j-1}})} \right)^2}{n^\nu \log\log n^\nu} \frac{n}{p} \leq \frac{C_9 \left(\sum_{j=1}^{p} j^{\frac{\nu-1}{2}} \right)^2}{n^\nu \log\log n^\nu} \frac{n}{p} \leq$$

$$\leq \frac{C_{10} \, p^{2\left(\frac{\nu-1}{2}+1\right)}}{n^\nu \log\log n^\nu} \frac{n}{p} \leq \frac{C_{11}}{\log^{1+\rho} n} \, .$$

Since by Chebyshev inequality for sufficiently large n

$$1 - \max_{0 \le i \le k-2} P\left(|S_{I_n} - S_{I_{(i+2)P}}| \ge \frac{b \chi(n)}{3}\right) \le \frac{1}{2}$$

we finish the proof.

__Theorem 7.4.3.__ Let ξ_t , $t \in \mathbb{Z}^\nu$, be a φ -mixing s.r.f. with mixing coefficient $\varphi_{m,\infty}(\tau)$ and let

1. $E\,\xi_0 = 0, \quad E\,|\xi_0|^{2+\delta} < \infty, \quad \delta > 0$

2. $\varphi_{m,\infty}(\tau) \le m^\tau \varphi(\tau), \quad \tau > 0$, and

$$\sum_{\tau=1}^{\infty} \varphi^{\frac{1}{2}}(\tau) < \infty.$$

Then $\quad \sigma^2 = \sum_{t \in \mathbb{Z}^\nu} E\,\xi_0 \xi_t < \infty.$

If in addition $\sigma^2 \ne 0$ then for law of the random field $\xi_t, t \in \mathbb{Z}^\nu$, the law of the iterated logarithm is valid.

Using Theorems 7.4.1, 7.3.2 and inequality 3.3.2 one can prove this theorem by the same way as the previous one.

8. DESCRIPTION OF RANDOM FIELDS BY MEANS OF CONDITIONAL

PROBABILITIES

The way of determining a random process by means of its conditional probabilities has been used in probability theory quite for a long time, for example, the definition of Markov chain with the help of its transition matrix. However, it was only in connection with the demands of statistical physics, in particular, with those of the strict definition of Gibbs random field the wide possibilities of this approach were cleared up. Below we will describe some facts of this theory using the results of fundamental works /29/ , /32/ . As it turned out in contrast to defining a random field by means of its finite-dimensional distributions the uniqueness of a random field with the given conditional distribution does not come from its existence here. As it will be shown in this Chapter for the uniqueness to hold here it is necessary to require that the field should possess some properties of decay of correlations, for example, those of mixing. In doing this it is possible to indicate the estimates of the mixing coefficient in terms of conditional probabilities. At the end of this chapter c.l.th. will be presented for the random field under some requirements for its conditional distribution.

8.1 **Existence of Random Fields** with Given Conditional Distribution

Let ξ_t, $t \in \mathbb{Z}^\nu$, be a random field, the components of which assume values in some compact metric space X . For any $V \subset \mathbb{Z}^\nu$ let

$$X_V = \left\{ (x_t, t \in V), \ x_t \in X \right\}, \quad X_{\mathbb{Z}^\nu} = \Omega$$

and if $x \in \Omega$, then x_V will denote its restriction on X_V.
Let $V_1, V_2 \in W$, where

$$W = \left\{ V \subset \mathbb{Z}^\nu : \ |V| < \infty \right\}, \ V_1 \cap V_2 = \varnothing .$$

For $x \in \Omega$ denote $(x_{V_1}, \ x_{V_2}) = x_{V_1 \cup V_2}$.
We will define a random field either by means of its distribution $\mathbb{P}$ on σ-algebra $\mathfrak{M}$ generated by cylindrical subsets or by means of finite-dimensional probabilities

$$\mathbb{P} = \left\{ P_V , \ V \in W \right\} .$$

Identically we will denote $\mathcal{B}_V$ both as the product

$$\mathcal{B}_V = \bigotimes_{t\in V} \mathcal{B}_t , \qquad \mathcal{B}_t \equiv \mathcal{B} \qquad (\ \mathcal{B} \ \text{is a} \ \sigma\text{-algebra of Borel}$$

subsets from X) and the σ-algebra of the cylindrical subsets with base $V\in W$. As a system of conditional probabilities we will call the system of functions

$$Q = \left\{ q_V \left({}^{B}\!/\!x_{V'} \right) , \quad V\in W, \quad V' = \mathbb{Z}^\nu \setminus V \right\} \qquad (8.1.1)$$

measurable with respect to $\quad x_{V'} = (x_t, \ t\in\mathbb{Z}^\nu\setminus V\}$ and being probability measures on $\mathcal{B}_V$.

Further we will consider only consistent systems of conditional probabilities, i.e. those that for all $V_1, V_2 \in W$, $V_1 \cap V_2 = \varnothing$,
$B_1 \in \mathcal{B}_{V_1}$, $\quad B_2 \in \mathcal{B}_{V_2}$ and $\quad \overline{x} \in X_{\mathbb{Z}^\nu \setminus (V_1 \cup V_2)}$
the following relations are valid:

$$q_{V_1 \cup V_2} \left({}^{B_1 \times B_2}\!/\!\overline{x} \right) = \int_{B_2} q_{V_1} \left({}^{B_1}\!/\!z\,\overline{x} \right) \left(q_{V_1 \cup V_2} \right)_{V_2} \left({}^{dz}\!/\!\overline{x} \right), \quad (8.1.2)$$

where

$$\left(q_{V_1 \cup V_2} \right)_{V_2} \left({}^{B_2}\!/\!\overline{x} \right) = q_{V_1 \cup V_2} \left({}^{B_2 \times X_{V_1}}\!/\!\overline{x} \right), \quad B_2 \in \mathcal{B}_{V_2} .$$

We will also assume that the conditional probabilities (8.1.1) are continuous in the product-topology with respect to
$\overline{x} = (x_t, \ t\in\mathbb{Z}^\nu\setminus V)$. It is said that a random field
$\xi_t, \ t\in\mathbb{Z}^\nu$, has (8.8.1) as its conditional distribution if with probability 1 for all $V\in W$ the next equality is valid:

$$P_z \left({}^{(\xi_t, \ t\in V)\in B}\!/\!\xi_t = x_t, \ t\in\mathbb{Z}^\nu\setminus V \right) = q_V \left({}^{B}\!/\!x_t, \ t\in\mathbb{Z}^\nu\setminus V \right).$$

Technically it will be quite useful to introduce a random field for each $V\in W$ and $x\in\Omega$ with the following distribution:

$$\mathbb{P}^{V, x}(B) = q_V \left({}^{B}\!/\!x_{\mathbb{Z}^\nu\setminus V} \right), \qquad (8.1.3)$$

$$\mathbb{P}^{V, x} \left(x_{\mathbb{Z}^\nu\setminus V} \right) = 1 .$$

Denote $C(\Omega)$ as the space of continuous function defined on Ω . It follows from the continuity of conditional probabilities that

$$\mathbb{P}^{V, x} f = \int f(u) \, \mathbb{P}^{V, x}(du) \in C(\Omega) , \quad x\in\Omega.$$

Note, that the consistency of conditional probabilities is equivalent to the following equality : for any $V_1, V_2 \in W$,

$V_1 \cap V_2 = \emptyset$ and any bounded continuous functions $f(x_{V_1})$, $g(x_{V_2})$

$$\iint\limits_{X_{V_1} \times X_{V_2}} f(x_{V_1}) g(x_{V_2}) q_{V_1 \cup V_2} \left(dx_{V_1} \times dx_{V_2} / \overline{x} \right) =$$

$$= \int\limits_{X_{V_2}} g(x_{V_2}) \left(\int\limits_{X_{V_1}} f(x_{V_1}) q_{V_1} \left(dx_{V_1} / \overline{x}, x_{V_2} \right) (q_{V_1 \cup V_2})_{V_2} \left(dx_{V_2} / \overline{x} \right) \right).$$

The latter can be rewritten in terms of distributions (8.1.3) in such way:

$$\int f(x_{V_1}) g(x_{V_2}) \mathbb{P}^{\overline{x}, V_1 \cup V_2}(dx) =$$

$$= \int g(x_{V_2}) \left(\int f(x_{V_1}) q_{V_1} \left(dx_{V_1} / x_{\mathbb{Z}^\nu \setminus V_1} \right) \right) \mathbb{P}^{\overline{x}, V_1 \cup V_2}(dx). \tag{8.1.4}$$

Denote by $G(Q)$ the set of random fields with given conditional distribution Q .

<u>Theorem 8.1.1</u> (/29/). The set $G(Q)$ is not empty. If the system Q is translation-invariant, then there exists a homogeneous random field with Q as a system of its conditional distribution.

<u>Proof</u>. Since X represents a metric compact space, then according to Tikhonov's theorem the same is with Ω . By virtue of this the space of the distributions of random fields $\mathcal{P}$ is complete and compact. Let $\ldots V_k \subset V_{k+1} \subset \ldots$, $k \in \mathbb{N}$, be some increasing sequence of finite subsets of $\mathbb{Z}^\nu$, such that

$\bigcup\limits_k V_k = \mathbb{Z}^\nu$. and let $\overline{x}_k \in X_{\mathbb{Z}^\nu \setminus V_k}$, $k \in \mathbb{N}$, be the corresponding boundary conditions. Consider the random fields with the following distributions (specifications):

$$\mathbb{P}_k = \mathbb{P}^{V_k, \overline{x}_k} , \quad k \in \mathbb{N}.$$

Since $\mathcal{P}$ is complete and compact, then from any sequence of distributions $\mathbb{P}_k$ it is possible to choose a converged subsequence. So without losing generality we can assume that

$$\lim_{k \to \infty} \mathbb{P}_k = \overline{\mathbb{P}}$$

and demonstrate that the random field   is the field with
given conditional distribution Q . For this it is sufficient
to show that for any finite and non-intersecting $V_1, V_2 \subset \mathbb{Z}^\nu$ and
for any continuous functions f and g the next relation
holds:

$$\int f(x_{V_1}) \, g(x_{V_2}) \, \overline{\mathbb{P}} \, (dx) =$$

$$= \int g(x_2)(\int f(x_{V_1}) \, q_{V_1} \, (^{dx_{V_1}}/x_{\mathbb{Z}^\nu \setminus V_1}) \, \overline{\mathbb{P}} \, (dx). \qquad (8.1.5)$$

For distributions $\mathbb{P}_k$ if k is sufficiently large we
have

$$\int f(x_{V_1}) \, g(x_{V_2}) \, \mathbb{P}_k \, (dx) =$$

$$= \int g(x_{V_2})(\int f(x_{V_1}) \, q_{V_1} \, (^{dx_{V_1}}/x_{\mathbb{Z}^\nu \setminus V_1}) \, \mathbb{P}_k \, (dx). \qquad (8.1.6)$$

By virtue of the continuity of subintegral functions in both
parts of (8.1.6) we may pass to the limit as $n \to \infty$ and thus
(8.1.5) is obtained. Hence $G(Q)$ is not empty. Now we will
prove the existence of the homogeneous distribution. Let V_k
be the cube with the side of length k and the center in the
origin. Let us consider the random fields with the following
distribution:

$$\overline{\mathbb{P}}^j = \left\{ \overline{P}^j_V, \; V \in W \right\}, \; j \in \mathbb{N},$$

$$\overline{P}^j_V \, (B) = \frac{1}{|V_j|} \sum_{t \in V_j} P_{V+t}(B), \quad B \in \mathcal{B}_V, \qquad (8.1.7)$$

where $\mathbb{P} = \left\{ P_V, \; V \in W \right\}$ is a random field with the given conditional distribution. It is easy to check that the linear com-
bination of random fields with given conditional distribution
Q also has Q as its conditional distribution. Now let
us choose a converged subsequence $\overline{\mathbb{P}}^{j_k}$, $j_k \to \infty$, and let $\overline{\mathbb{P}}$
be its limit. Then $\overline{\mathbb{P}}$ has the system Q as its condi-
tional distribution and, moreover, for any $V \in W$ $t_0 \in \mathbb{Z}^\nu$
and $B \in \mathcal{B}_V$ we have

$$|\overline{P}^j_V \, (B) - \overline{P}^j_{V+t_0}(B)| \leq \frac{|V_j \triangle (V_j + t_0)|}{|V_j|}, \qquad (8.1.8)$$

182

where Δ is a symmetric difference.

Now it is evident that the right-hand side in (8.1.8) tends to zero for fixed t_0 when $j \to \infty$ and it implies the homogeneity of random field $\mathbb{P}$.

8.2 Uniqueness of Random Fields with Given Conditional Distribution

In this section and in the next one we follow the articles /29/, /54/ and /75/.

Let $Q_i = \{ q_V^i , V \in W \}$, $i = 1, 2$ be two systems of conditional probabilities. Put

$$\rho_{a\beta} = \tfrac{1}{2} \sup_{i=1,2} \max \left\{ \rho\left(q_a^{(i)}(\cdot/x), q_a^{(i)}(\cdot/\tilde{x}) \right) \right\},$$

where sup takes over all $x, \tilde{x} \in X_{\mathbb{Z}^\nu \setminus \{a\}}$ such that $x = \tilde{x}$ everywhere except the point β, $\beta \in \mathbb{Z}^\nu$, $a \neq \beta$, $\rho_{a,a} = 0$. Here $\rho(\cdot, \cdot)$ is a variation distance between the corresponding distributions. Denote

$$\beta_a = \tfrac{1}{2} \sup \left\{ \rho\left(q_a^{(1)}(\cdot/x), q_a^{(2)}(\cdot/x) \right); \quad x \in X_{\mathbb{Z}^\nu \setminus a} \right\}.$$

Let Γ be an infinite matrix with elements $(\rho_{a\beta})_{a, \beta \in \mathbb{Z}^\nu}$ and Γ_V be its restriction on $a, \beta \in V$.
Denote

$$\chi_{a\beta}^V = \sum_{n=0}^{\infty} (\Gamma_V^n)_{a\beta}, \qquad a, \beta \in V \in W,$$

$$\chi_{a\beta} = \sum_{n=0}^{\infty} (\Gamma^n)_{a\beta}.$$

For any function $f \in C(\Omega)$ and $t \in \mathbb{Z}^\nu$ put

$$\tau_t(f) = \sup \left\{ |f(x) - f(\tilde{x})| \right\},$$

where sup extends over all $x, \tilde{x} \in \Omega$, such that $x = \tilde{x}$ except the point t, $t \in \mathbb{Z}^\nu$. The following inequality is evident:

$$|f(x) - f(\tilde{x})| \leq \sum_{t \in \mathbb{Z}^\nu} \tau_t(f), \qquad x, \tilde{x} \in \Omega . \tag{8.2.1}$$

$\underline{\text{Theorem 8.2.1 } (/75/)}.$ Let $\mathbb{P}^{(i)}$, $i = 1,2$, be the random fields with given conditional distributions $Q_i = \{ q_V^{(i)}, \ V \in W \}$ $i = 1,2$ respectively. If

$$\sum_{a \in \mathbb{Z}^\nu} \rho_{a\beta} \leqslant \alpha < 1 \qquad\qquad (8.2.2)$$

then for any function $f \in C(\Omega)$ the relation

$$\left| \int f(u) \, \mathbb{P}^{(1)}(du) - \int f(u) \, \mathbb{P}^{(2)}(du) \right| \leqslant \qquad\qquad (8.2.3)$$

$$\leqslant \sum_{a,\beta} \beta_\beta \, \chi_{a\beta} \, \tau_\beta (f)$$

is fulfilled.

Before proving this theorem let us give some of its consequences.

$\underline{\text{Theorem 8.2.2 } (/29/)}.$ Suppose that

$$\sum_{a \in \mathbb{Z}^\nu} \rho_{a\beta} \leqslant \alpha < 1$$

then there exists the unique random field with the given conditional distribution.

$\underline{\text{Proof}}.$ Let $\mathbb{P}^{(1)}$ and $\mathbb{P}^{(2)}$ have the same conditional distribution. Then $\beta_a = 0$ for all $a \in \mathbb{Z}^\nu$ and hence from (8.2.3) it follows:

$$\int f(u) \, \mathbb{P}^{(1)}(du) = \int f(u) \, \mathbb{P}^{(2)}(du)$$

for any $f \in C(\Omega)$. Hence $\mathbb{P}^{(1)} = \mathbb{P}^{(2)}$.

$\underline{\text{Theorem 8.2.3}(/29/)}.$ Let

$$\sum_{a \in \mathbb{Z}^\nu} \rho_{a\beta} \leqslant \alpha < 1$$

and $\mathbb{P}$ be the unique random field corresponding to the given conditional distribution $Q = \{ q_V, \ V \in W)$. Then for all V_1 , $V_2 \in W$, such that $V_1 \subset V_2$ the following relation holds:

$$\varphi (\mathcal{B}_{V_1} , \mathcal{B}_{V_2}) \leqslant \sum_{\beta \in V_1} \Big(\sum_{c \in V_2} \sum_{a \notin V_2} \rho_{ca} \, \chi_{a\beta} \Big). \qquad\qquad (8.2.4)$$

Proof. Let $\mathbb{P}^{V,x}$, $V \in W$, be a random field of type (8.1.3) associated with the system $Q = \{q_V, V \in W\}$. It is evident that its conditional distribution on V coincides with the system $\{q_{\widetilde{V}}, \widetilde{V} \subset V\}$. Hence, using Theorem 8.2.1 for all $x, \widetilde{x} \in X_{\mathbb{Z}^\nu \setminus V}$ we can write

$$|\mathbb{P}^{V,x} f - \mathbb{P}^{V,\widetilde{x}} f| = |\int f(u x) q_V (du/x) -$$

$$- \int f(u \widetilde{x}) q_V (du/\widetilde{x})| \le |\int f(u x) q_V (du/x) -$$

$$- \int f(u x) q_V (du/\widetilde{x})| + |\int (f(u x) - f(u \widetilde{x})) q_V (du/\widetilde{x})| \le$$

$$\le \sum_{a \in V, \, \beta \in V} \beta_a \, \chi_{a\beta}^V \, \tau_\beta(f) + \sum_{c \notin V} \tau_c(f).$$

By virtue of the inequality

$$\beta_a \le \sum_{c \notin V} \rho_{ca}$$

we obtain

$$|\mathbb{P}^{V,x} f - \mathbb{P}^{V,\widetilde{x}} f| \le \sum_{a \in V, \, \beta \in V} (\sum_{c \notin V} \rho_{ca}) \, \chi_{a\beta}^V \, \tau_\beta(f) + \sum_{c \notin V} \tau_c(f). \tag{8.2.5}$$

Suppose that $A \in \mathcal{B}_{V_1}$ is an open set and I_A is its indicator. Approximating I_A by the appropriate sequence of continuous functions from (8.2.5) we have

$$|q_V (A/x) - q_V (A/\widetilde{x})| \le \sum_{a \in V_2, \, \beta \in V_1} (\sum_{c \notin V_2} \rho_{ca}) \, \chi_{a\beta}^{V_2}. \tag{8.2.6}$$

Thus for $B \in \mathcal{B}_{\mathbb{Z}^\nu \setminus V_2}$ it follows:

$$|\varphi(A, B)| = |\int I_B (x) q_V (A/x) \, \mathbb{P}(dx) -$$

$$- \iint I_B (x_1) q_V (A/x_2) \, \mathbb{P}(dx_1) \, \mathbb{P}(dx_2)| \le$$

$$\leqslant \mathbb{P}(B) \sum_{b \in V_1, a \in V_2} \Big(\sum_{c \notin V_2} \rho_{ca} \Big) \chi_{ab}^{V_2} \leqslant$$

$$\leqslant \mathbb{P}(B) \sum_{b \in V_1} \sum_{a \in V_2} \Big(\sum_{c \notin V_2} \rho_{ca} \Big) \chi_{ab}.$$

The proof is finished.

Note that $\varphi\big(\mathcal{B}_{V_1}, \mathcal{B}_{\mathbb{Z}^\nu \setminus V_2}\big) \to 0$ if V_1 is fixed and $V_2 \uparrow \mathbb{Z}^\nu$. Really, if $a \in \mathbb{Z}^\nu$ is fixed, then $\sum_{c \notin V_2} \rho_{ca} \to 0$ as $V_2 \uparrow \mathbb{Z}^\nu$, and for fixed $b \in V_1$ we have

$$\sum_{a \in V_2} \sum_{c \notin V_2} \rho_{ca} \chi_{ab} \leqslant \alpha \sum_a \chi_{ab} \leqslant \alpha (1-\alpha)^{-1}.$$

Now we give the proof of Theorem 8.2.1.

We will say that function $(\alpha_a)_{a \in \mathbb{Z}^\nu}$ is an estimate for the random fields $\mathbb{P}^{(1)}$ and $\mathbb{P}^{(2)}$ if

$$|\mathbb{P}_1 f - \mathbb{P}_2 f| \leqslant \sum_{a \in \mathbb{Z}^\nu} \alpha_a \, r_a(f), \qquad f \in C(\Omega). \qquad (8.2.7)$$

For any estimate $(\alpha_a)_{a \in \mathbb{Z}^\nu}$ let us define the estimate $(\hat{\alpha}_a)_{a \in \mathbb{Z}^\nu}$ with the help of the equality:

$$\hat{\alpha}_a(b) = \begin{cases} \alpha_a, & a \neq b \\[2mm] \beta_b + \sum_{c \neq b} \alpha_c \, \rho_{cb}, & a = b. \end{cases}$$

<u>Lemma 8.2.1.</u> If $(\alpha_a)_{a \in \mathbb{Z}^\nu}$ is an estimate for $\mathbb{P}^{(1)}$ and $\mathbb{P}^{(2)}$ then for any $b \in \mathbb{Z}^\nu$, $\hat{\alpha}_a(b)$, $a \in \mathbb{Z}^\nu$, is also the estimate for $\mathbb{P}^{(1)}$ and $\mathbb{P}^{(2)}$.

<u>Proof.</u> By virtue of both the definition of conditional probabilities and specifications we have

$$|\mathbb{P}^{(1)} f - \mathbb{P}^{(2)} f| = |\mathbb{P}^{(1)}(\mathbb{P}_1^b f) - \mathbb{P}^{(2)}(\mathbb{P}_2^b f)|,$$

where $\mathbb{P}_i^{b,x}$, $i = 1,2$ are the specifications constructed in the point $b \in \mathbb{Z}^\nu$ according to distributions $\mathbb{P}^{(i)}$,

$$\mathbb{P}^{(i)} f = \int f(w) \, \mathbb{P}^{(i)}(dw), \qquad i = 1,2.$$

Hence

$$|P^{(1)}f - P^{(2)}f| \leq |P^{(1)}(P_1^{\theta}f) - P^{(1)}(P_2^{\theta}f)| +$$
$$+ |P^{(1)}(P_2^{\theta}f) - P^{(2)}(P_2^{\theta}f)| . \qquad (8.2.8)$$

By the fact that $(\alpha_a)_{a \in \mathbb{Z}^\nu}$ is the estimate for $P^{(1)}$ and $P^{(2)}$ we have

$$|P^{(1)}f - P^{(2)}f| \leq |P^{(1)}(P_1^{\theta}f) - P^{(1)}(P_2^{\theta}f)| + \sum_{a \in \mathbb{Z}^\nu} \alpha_a \tau_a (P_2^{\theta}f) .$$

Now if $f \in C(\Omega)$ and $x, \tilde{x}$ are two configurations from Ω which coincide, everywhere, exept the point $a \in \mathbb{Z}^\nu$, then

$$|(P_1^{\theta}f)(x) - (P_2^{\theta}f)(\tilde{x})| = |\int f(xu) P_1^{\theta}(du/x) -$$
$$- \int f(\tilde{x}u) P_2^{\theta}(du/\tilde{x})| = |\int (f(xu) - f(xv))(P_1^{\theta}(du/x) - P_2^{\theta}(du/x))| \leq$$
$$\leq \tau_\theta (f) \beta_\theta , \qquad v \in X_\theta , \quad \theta \in \mathbb{Z}^\nu$$

Hence
$$\|P^{(1)}f - P^{(2)}f\| \leq \tau_\theta (f) \beta_\theta + \sum_{a \in \mathbb{Z}^\nu} \alpha_a \tau_a (P_2^{\theta}f) . \qquad (8.2.9)$$

Further

$$|(P_2^{\theta}f)(x) - (P_2^{\theta}f)(\tilde{x})| =$$
$$= |\int f(xu) P_2^{\theta}(du/x) - \int f(\tilde{x}u) P_2^{\theta}(du/\tilde{x})| \leq$$
$$\leq |\int (f(xu) - f(\tilde{x}u)) P_2^{\theta}(du/x)| +$$
$$+ |\int f(\tilde{x}u)(P_2^{\theta}(du/x) - P_2^{\theta}(du/\tilde{x}))| \leq$$
$$\leq \tau_a (f) + \rho_{a\theta} \tau_\theta (f) .$$

Hence
$$\tau_a(P_2^{\theta}f) \leq \tau_a (f) + \rho_{a\theta} \tau_\theta (f) , \quad a \neq \theta$$
and

$$\rho_a(\mathbb{P}_2^{\beta} f) = 0, \qquad a \neq \beta.$$

Using (8.2.10) we can continue inequality (8.2.9). We have

$$|\mathbb{P}^{(1)} f - \mathbb{P}^{(2)} f| \leq \beta_\beta \, \tau_\beta(f) + \sum_{a \neq \beta} \alpha_a \, \tau_a(f) +$$

$$+ \sum_{a \neq \beta} \alpha_a \, \rho_{a\beta} \, \tau_\beta(f) = \left(\beta_\beta + \sum_{a \neq \beta} \alpha_a \rho_{a\beta}\right) \tau_\beta(f) + \sum_{a \neq \beta} \alpha_a \tau_a(f).$$

The lemma is proven.

For further study let us check that $\alpha_a \equiv 1$, $a \in \mathbb{Z}^{\nu}$, is the estimate for $\mathbb{P}^{(1)}$ and $\mathbb{P}^{(2)}$, i.e.,

$$|\mathbb{P}^{(1)} f - \mathbb{P}^{(2)} f| \leq \sum_{a \in \mathbb{Z}^{\nu}} \tau_a(f).$$

Really, we have

$$|\mathbb{P}^{(1)} f - \mathbb{P}^{(2)} f| \leq \sup_u |f(u) - \mathrm{const}| \, \rho(\mathbb{P}^{(1)}, \mathbb{P}^{(2)}) \leq$$

$$\leq \sup |f(u) - \mathrm{const}| \leq \sum_{a \in \mathbb{Z}^{\nu}} \tau_a(f), \qquad \mathrm{const} = f(u^0), \quad n^0 \in \Omega.$$

Now let's consider the function

$$\alpha_a^{\tau} = \begin{cases} 1, & a \notin V \\ \tau \sum_{c \in V} \beta_c \, \chi_{ca}^{\nu} + \tau \sum_{\beta \notin V, \, c \in V} \rho_{\beta c} \, \chi_{ca}^{\nu}, & a \in V. \end{cases}$$

Lemma 8.2.2(/75/). If $\tau \beta_a \geq 1$ for all $a \in V$, then α_a^{τ} is the estimate for $\mathbb{P}^{(1)}$ and $\mathbb{P}^{(2)}$.

Proof. Since $\alpha_a \equiv 1$, $a \in \mathbb{Z}^{\nu}$, is an estimate it is sufficient to show that under conditions of Lemma 8.2.2 $\alpha_a^{\tau} \geq 1$ for all $a \in V$. We have

$$\tau \sum_{c \in V} \beta_c \, \chi_{ca}^{\nu} = \tau \beta_a \, \chi_{aa}^{\nu} + \tau \sum_{c \in V \setminus \{a\}} \beta_c \, \chi_{ca}^{\nu} \geq \tau \beta_a \geq 1.$$

Here we took advantage of the fact that

$$\chi_{aa}^{V} = \sum_{n=0}^{\infty} \left(\Gamma_{V}^{n}\right)_{aa} = 1 .$$

Further we can write

$$\sum_{a\in V} \chi_{ca}^{V} \, \rho_{a\beta} = \sum_{a\in V} \sum_{n=0}^{\infty} \left(\Gamma_{V}^{n}\right)_{ca} \, \rho_{a\beta} =$$

$$= \sum_{n=0}^{\infty} \sum_{a\in V} \left(\Gamma_{V}^{n}\right)_{ca} \, \rho_{a\beta} = \sum_{n=0}^{\infty} \left(\Gamma_{V}^{n+1}\right)_{c\beta} =$$

$$= \sum_{n=1}^{\infty} \left(\Gamma_{V}^{n}\right)_{c\beta} = \sum_{n=0}^{\infty} \left(\Gamma_{V}^{(n)}\right)_{c\beta} - \left(\Gamma_{V}^{0}\right)_{c\beta} = \chi_{c\beta}^{V} - \delta_{c\beta} .$$

Hence

$$\sum_{a\in\mathbb{Z}^{V}} \alpha_{a}^{\tau} \, \rho_{a\beta} = \sum_{a\in V} \left(\tau \sum_{c\in V} \beta_{c} \, \chi_{ca}^{V} + \tau \sum_{a\notin V,\, c\in V} \rho_{dc} \, \chi_{ca}^{V}\right) \rho_{a\beta} +$$

$$+ \sum_{a\notin V} \rho_{a\beta} =$$

$$= \sum_{a\notin V} \rho_{a\beta} + \tau \sum_{c\in V} \beta_{c} \left(\chi_{c\beta}^{V} - \delta_{c\beta}\right) + \tau \sum_{d\notin V,\, c\in V} \rho_{dc} \, \chi_{c\beta}^{V} -$$

$$- \tau \sum_{d\notin V} \rho_{d\beta} =$$

$$= \sum_{a\notin V} \rho_{a\beta} + \tau \sum_{c\in V} \beta_{c} \, \chi_{c\beta}^{V} - \tau \beta_{\beta} + \tau \sum_{d\notin V,\, c\in V} \rho_{dc} \, \chi_{c\beta}^{V} -$$

$$- \tau \sum_{d\notin V} \rho_{d\beta} =$$

$$= \alpha_{\beta}^{\tau} - \tau \beta_{\beta} - (\tau - 1) \sum_{d\notin V} \rho_{d\beta} .$$

Let
$$\bar{\tau} = \inf\left\{\tau \in R^1 : \left(\alpha_a^{\tau}\right)_{a \in \mathbb{Z}^{\nu}}\right\} \quad \text{be an estimate for}$$
$$\mathbb{P}^{(1)}, \mathbb{P}^{(2)}.$$

Assuming that $\bar{\tau} > 1$ we will lead to a contradiction and hence Theorem 8.2.1 will be proven. Consider the estimate such that

$$\alpha_a \leq \alpha_a^{\bar{\tau}(1+\delta)}, \qquad a \in \mathbb{Z}^{\nu}, \quad \delta > 0.$$

The estimate of this type always exists, for example,

$$\alpha_a \equiv 1.$$

Then from definition of $\bar{\tau}$ there exists a point $\beta \in V$ such that

$$\alpha_\beta > \alpha_\beta^{\bar{\tau}(1-\delta)}.$$

Now for sufficiently small $\delta > 0$ we can write

$$\hat{\alpha}_a(\beta) = \beta_\beta + \sum_{a \neq \beta} \alpha_a \rho_{a\beta} \leq \beta_\beta + \sum_{a \in \mathbb{Z}^{\nu}} \alpha_a^{\bar{\tau}(1+\delta)} \rho_{a\beta} \leq$$

$$\leq \alpha_\beta^{\bar{\tau}(1+\delta)} - \left(\bar{\tau}(1+\delta)-1\right)\left(\beta_\beta + \sum_{d \notin V} \rho_{d\beta}\right) =$$

$$= \alpha_\beta^{\bar{\tau}(1-\delta)} + 2\bar{\tau}\delta\alpha_\beta^{1} - \left(\bar{\tau}(1+\delta)-1\right)\left(\beta_\beta + \sum_{d \notin V} \rho_{d\beta}\right) \leq$$

$$\leq \alpha_\beta^{\bar{\tau}(1-\delta)}.$$

Repeating the same conclusions we obtain that

$$\alpha_a \leq \alpha_a^{\bar{\tau}(1-\delta)}$$

for all $a \in V$ and, hence, for all $a \in \mathbb{Z}^{\nu}$. It contradicts the definition of $\bar{\tau}$. Thus $\bar{\tau} \leq 1$ and the proof is finished.

8.3 Decay of Correlation and Central Limit Theorem

It is convenient to formulate the decay of correlation for the random fields with the given conditional distribution by means of some semimetric d , i.e., by non-negative symmetric function for which the triangle inequality is valid.

Proposition 8.3.1. Let $\mathbb{P}$ be a random field with the given conditional distribution, such that for some semimetric d the following inequality holds:

$$\sum_{a} \rho_{a\beta} \exp \left\{ d(a,\beta) \right\} \leq \alpha < 1 . \tag{8.3.1}$$

Then for any $V, \widetilde{V} \in W, \ V \subset \widetilde{V},$

$$\varphi(\mathcal{B}_V, \mathcal{B}_{\mathbb{Z}^\nu \setminus \widetilde{V}}) \leq \exp \left\{ -d(V, \mathbb{Z}^\nu \setminus \widetilde{V}) \right\} |V| \, \alpha \, (1-\alpha)^{-1}.$$

Proof. According to Theorem 8.2.3 we have

$$\varphi(\mathcal{B}_V, \mathcal{B}_{\mathbb{Z}^\nu \setminus \widetilde{V}}) \leq \sum_{\beta \in V} \left(\sum_{c \notin \widetilde{V}, \, a \in \widetilde{V}} \rho_{ca} \chi_{a\beta} \right).$$

For any $\beta \in V, \ a \in \widetilde{V}, \ c \notin \widetilde{V}$ we can write

$$\exp \left\{ d(V, \mathbb{Z}^\nu \setminus \widetilde{V}) \right\} \leq \exp \left\{ d(\beta, c) \right\} \leq$$

$$\leq \exp \left\{ d(\beta, a) \right\} \exp \left\{ d(a, c) \right\} .$$

Hence

$$\varphi(\mathcal{B}_V, \mathcal{B}_{\mathbb{Z}^\nu \setminus \widetilde{V}}) \leq$$

$$\leq \exp \left\{ -d(V, \mathbb{Z}^\nu \setminus \widetilde{V}) \right\} \sum_{\beta \in V} \left(\sum_{c \notin \widetilde{V}, \, a \in \widetilde{V}} \rho_{ca} \exp \left\{ d(c,a) \right\} \chi_{a\beta} \exp \left\{ d(a,\beta) \right\} \right) \leq$$

$$\leq \exp \left\{ -d(V, \mathbb{Z}^\nu \setminus V) \right\} |V| \, \alpha \, \sup \left\{ \sum_{a} \chi_{a\beta} \exp \left\{ d(a,\beta) \right\} \right\}, \quad \beta \in \mathbb{Z}^\nu .$$

From the triangle inequality and (8.3.1) it follows that

$$\sum_{a} \chi_{a\beta} \exp \left\{ d(a,\beta) \right\} \leq (1-\alpha)^{-1} .$$

Proposition 8.3.2 is proven.

Commenting condition 8.3.1 note that if $d(a,\beta) = \varepsilon |a - \beta|$, $\varepsilon > 0$, then it is equivalent to the exponential decay of correlations and if

$$d(a,b) = d_{\varepsilon,\gamma}(a,b) = \min(\varepsilon|a-b|, \ \gamma \log(1+|a-b|))$$

the decay of the correlation is power one.
Note that for the homogeneous system of conditional probabilities
$\rho_{ab} = \rho_{a-b\,0}$ and hence Condition 8.3.1 may be rewritten in such way

$$\sum_{a\in\mathbb{Z}^\nu} \rho_{a0} \exp\{d(0,a)\} \le \alpha < 1.$$

Theorem 8.3.1. Let a centered stationary random field ξ_t , $t\in\mathbb{Z}^\nu$, taking values in some compact space $X \subset R^1$ have a consistent and continuous system of conditional distributions and $E\xi_0^2 < \infty$
Suppose also that

$$\sum_{a\in\mathbb{Z}^\nu} \rho_{a0} < 1$$

and for $\gamma > 2\nu$,

$$\sum_{a\in\mathbb{Z}^\nu} \rho_{a0} \exp\{d_{\varepsilon,\gamma}(a,0)\} \le \alpha < 1.$$

Then the series

$$\sigma^2 = \sum_{t\in\mathbb{Z}^\nu} E\xi_0\xi_t$$

converges and if $\sigma^2 \neq 0$, then for this random fields c.l.th. holds.

This theorem is the consequence of the received results and Theorem 7.2.2.

9. GIBBS RANDOM FIELDS

The concept of Gibbs random field arisen in statistical physics with a view to describe a physical system in infinite volume and first was given by Dobrushin /29/ ,/30/ , Lanford and Ruelle /76/ . Gibbs random fields are those probability measures which have the given conditional distributions. These conditional probabilities have a specific form involving potential. In this chapter we state the basic facts of the theory of Gibbs random fields as well as some applications of the results of previous chapters.

9.1 Existence and Uniqueness of Gibbs Random Fields, Weak Dependence of Components

Let X be a compact metric space with the σ-algebra of its Borel subsets $\mathcal{F}$ and a finite measure μ , such that $\mu(X) > 0$. For any $I \subseteq \mathbb{Z}^\nu$ let's introduce a measurable space

$$L_I = \bigcup_{J \in W(I)} X_J ,$$

$$X_J = \bigotimes_{t \in J} X_t , \quad X_t = X, \quad W(I) = \left\{ J \subset I, \quad |J| < \infty \right\}$$

$$W(\mathbb{Z}^\nu) = W, \quad L_{\mathbb{Z}^\nu} = L$$

with σ-algebra $\mathcal{M}$ and the measure m the restrictions of which on each X_I, $I \in W$, coincide with

$$\mathcal{F}_I = \bigotimes_{t \in I} \mathcal{F}_t , \quad \mathcal{F}_t = \mathcal{F}, \quad \mu_I = \bigotimes_{t \in I} \mu_t , \quad \mu_t = \mu,$$

respectively.

For any $I \in W$ and $x \in X_I$ let's define a shift τ_t, $t \in \mathbb{Z}^\nu$, of the element x by the following way

$$\tau_t(x) = \left\{ \tilde{x}_s, \ s \in I + t \right\} \in X_{I+t} , \quad \tilde{x}_s = x_{s-t}, \quad s \in I + t.$$

A function Φ defined on L and invariant with respect to all shifts τ_t, $t \in \mathbb{Z}^\nu$, we shall call a potential. Further we shall consider potentials Φ such that

$$\| \Phi \| = \sum_{J : 0 \in J \in W} |J| \sup_{x \in X_J} | \Phi(x)| < \infty. \qquad (9.1.1)$$

The set $\mathcal{B}$ of potentials Φ satisfying (9.1.1) forms a Banach space with the norm defined by (9.1.1). For any $I \in W$

and $\bar{x} \in X_{\mathbb{Z}^\nu \setminus I}$ let's introduce a potential energy for configuration $x \in X_I$ and boundary conditions $\bar{x} \in X_{\mathbb{Z}^\nu \setminus I}$

$$U_I^{\bar{x}, \Phi}(x) = U_I^{\bar{x}}(x) = \sum_{J: \emptyset \neq J \subset I} \sum_{\bar{J} \in W(\mathbb{Z}^\nu \setminus I)} \Phi(x_J, \bar{x}_{\bar{J}}). \qquad (9.1.2)$$

A probability distribution $\Gamma_I^{\bar{x}, \Phi}$ on $\mathcal{F}_I$ we shall call Gibbs distribution on finite volume I with boundary conditions $\bar{x}$ if it is absolutely continuous with respect to measure μ_I and has the density in the form

$$q_I^{\bar{x}, \Phi}(x) = q_I^{\bar{x}}(x) = Z_I^{-1}(\bar{x}) \exp\left\{- U_I^{\bar{x}}(x)\right\}, \quad I \in W, \qquad (9.1.3)$$

$$Z_I^{-1}(\bar{x}) = Z_I^{-1}(\bar{x}, \Phi) = \int_{X_I} \exp\left\{- U_I^{\bar{x}}(x)\right\} \mu_I(dx).$$

The fundamental property of the system

$$\left\{ q_I^{\bar{x}}(x), \ I \in W \right\}$$

is its consistency in the sense of conditional distributions. Really, it is sufficient to check that for any $V_1, V_2 \in W$

$$q_{V_1 \cup V_2}^{\bar{x}}(xz) = q_{V_1}^{z\bar{x}}(x) \int_{X_{V_1}} q_{V_1 \cup V_2}^{\bar{x}}(yz) \mu_{V_1}(dy), \quad x \in X_{V_1},$$

$$z \in X_{V_2}, \ \bar{x} \in X_{\mathbb{Z}^\nu \setminus (V_1 \cup V_2)}.$$

Since

$$U_{V_1 \cup V_2}^{\bar{x}}(xz) = U_{V_2}^{\bar{x}}(z) + U_{V_1}^{z\bar{x}}(x)$$

then

$$\frac{q_{V_1 \cup V_2}^{\bar{x}}(xz)}{\int_{X_{V_1}} q_{V_1 \cup V_2}^{\bar{x}}(zy) \mu_{V_1}(dy)} =$$

$$= \frac{\exp\left\{- U_{V_2}^{\bar{x}}(z)\right\} \exp\left\{- U_{V_1}^{z\bar{x}}(x)\right\}}{\exp\left\{- U_{V_2}^{\bar{x}}(z)\right\} \int_{X_{V_1}} \exp\left\{- U_{V_1}^{z\bar{x}}(y)\right\} \mu_{V_1}(dy)} = q_{V_1}^{z\bar{x}}(x).$$

Note also that according to the definitions (9.1.2), (9.1.3) and
the condition (9.1.1) for arbitrary $\varepsilon > 0$ there exists a number
τ_ε such that for any two boundary conditions x' and x'' which
satisfy the relation

$$x'(t) = x''(t) , \qquad |t| \leqslant \tau_\varepsilon$$

the following inequality holds

$$\int_{X_I} | q_I^{x'}(x) - q_I^{x''}(x) | \, \mu_I(dx) < \varepsilon .$$

The latter implies the continuity of the given system of conditional
distributions (9.1.3) in product topology. Thus one can apply the
general results of Chapter 8 to the system (9.1.3).

A random field and its distribution we shall call Gibbsian if its
conditional distribution is given by the relations (9.1.3). Theorem
8.1.1 shows that the set of Gibbs random fields corresponding to the
potential (9.1.1) is not empty. Moreover, since the proof of Theorem
8.1.1 is constructive we can formulate another definition of the Gibbs
random field:

Suppose that for some sequence of increasing subsets Λ_n, $n \in \mathbb{N}$
$\bigcup_n \Lambda_n = \mathbb{Z}^\nu, \cdots \Lambda_n^C \Lambda_{n+1}^C \cdots$ and boundary conditions $\bar{x}_n \in \mathbb{Z}^\nu \backslash \Lambda_n$ the limit

$$\lim_{n \to \infty} (q_{\Lambda_n}^{\bar{x}_n})_I (x) = Q_I (x)$$

exists.

It is not difficult to verify that the system of distribution functions $\{ Q_I, \ I \in w \}$ is consistent in a sense of Kolmogorov and hence
it determines the random field which we will call limiting Gibbs random field. Let us consider the Banach space R of all real bounded
measurable functions $R = \{ \tau_I(x), x \in X_I, \ I \in W \}$

$$\| \tau \| = \sup_{I \in W} 2 |I|^{-1} \int_{X_I} | \tau_I(x) | \, \mu_I(dx)$$

<u>Theorem 9.1.0 (/30/)</u>. Let Λ_n, $n \in \mathbb{N}$, be a sequence of finite subsets
such that $\bigcup_n \Lambda_n = \mathbb{Z}^\nu$. Let $\mathcal{U}_n$ be the closed convex span of
Gibbs distributions in finite volume Λ_n with different boundary
conditions. Then

$$\mathcal{U}_{n+1} \subset \mathcal{U}_n , \quad n \in \mathbb{N}$$

and the intersection of the sets $\mathcal{U}_n$, $n \in \mathbb{N}$ coincides with the family of Gibbs distributions. This set represents a non-empty closed

195

convex span of the limiting Gibbs distributions.

Note also that Gibbs random fields represent some specific way
to   describe   general random fields (see,for example,/4/,
/5/,/117/).

__Theorem 9.1.1.__   Suppose the potential $\Phi \in \mathfrak{B}$ satisfies the
following condition

$$\frac{1}{2} e^{4\|\Phi\|} (e^{4\|\Phi\|} - 1) < 1 . \tag{9.1.4}$$

Then the corresponding Gibbs random field exists, is unique
and mixing with the coefficient

$$\varphi(\mathcal{F}_{V_1}, \mathcal{F}_{V_2}) \leq C_\Phi \sum_{\gamma \in G(V_1,V_2)} \tau(t_2,t_1)\tau(t_3,t_2)\ldots \tau(t_{K_\gamma}, t_{K_{\gamma-1}}),$$

where $G(V_1,V_2)$ is the set of all paths connecting V_1 and
V_2 , i.e.,the set of all sequences $\{t_1,\ldots,t_K\}$ such that

$t_1 \in V_1$, $t_K \in V_2$, $0 < C_\Phi < \infty$ and

$$\tau(a,b) = \exp\left\{4 \sum_{J:\, a,b \in J \in W} \sup_{x \in X_J} |\Phi(x)|\right\} - 1 . \tag{9.1.5}$$

__Proof__   We have   (see 8.2.3)

$$\rho_{ab} = \frac{1}{2} \sup_{x',x''} \int_{X_a} \left| \frac{\exp\{-\mathcal{U}_a^{x'}(x)\}}{Z_a(x')} - \frac{\exp\{-\mathcal{U}_a^{x''}(x)\}}{Z_a(x'')} \right| \mu_a(dx),$$

where $x', x'' \in X_{\mathbb{Z}^\nu \setminus \{a\}}$ are different only in the point
$b \in \mathbb{Z}^\nu$.

Further

$$\rho_{ab} \leq \frac{1}{2} \frac{e^{2\|\Phi\|}}{\mu^2(X)} \sup_{x',x''} \int\int_{X_a X_a} \left(\exp\{-\mathcal{U}_a^{x'}(x) - \mathcal{U}_a^{x''}(y)\} - \right.$$

$$\left. - \exp\{-\mathcal{U}_a^{x''}(x) - \mathcal{U}_a^{x'}(y)\}\right) \mu_a(dx)\mu_a(dy) \leq$$

$$\leq \frac{1}{2}\frac{e^{4\|\Phi\|}}{\mu^2(X)} \sup_{x',x''} \iint_{X_a X_a} \left| \exp\left\{ -U_a^{x'}(x) + U_a^{x''}(x) - U_a^{x''}(y) + U_a^{x'}(y) \right\} - \right.$$

$$\left. - 1 \right| \mu_a(dx)\,\mu_a(dy) \leq$$

$$\leq \frac{1}{2}e^{4\|\Phi\|}\left(\exp\left\{ 4 \sum_{J:\,a,\beta\in J\in W} \sup_{x\in X_J} |\Phi(x)| \right\} - 1\right) = \frac{1}{2}e^{4\|\Phi\|}\,\tau_{a\beta}.$$

Whence taking into account the super-additivity of the function $e^x - 1$, $x > 0$ (i.e. the property $e^x - 1 + e^y - 1 < e^{x+y} - 1$, $x, y > 0$) we have

$$\sum_{\beta\in\mathbb{Z}^\nu\setminus\{a\}} \rho_{a\beta} \leq \frac{1}{2}e^{4\|\Phi\|}\left(e^{4\|\Phi\|} - 1\right) < 1.$$

Now applying Theorem 8.2.2 and 8.2.3 we finish the proof.

Sufficiently large and interesting class of the potentials is the class of so-called vacuum potentials. Suppose that the space X includes some element θ ("vacuum") such that $\mu(\theta) = 1$. Potential Φ is called vacuum potential if $\Phi(x) = 0$ for any $x \in X_I$, $I \in W$ such that $x_t = \theta$ at least for one $t \in I$. As it will be shown all the results obtained for general potentials are valid. Also for vacuum potentials with the norm

$$\|\Phi\| = \sum_{0\in J\in W} \sup_{x\in X_J} |\Phi(x)| < \infty. \tag{9.1.6}$$

Potential Φ is called finite if the number of subsets J such that $0 \in J$ and $\Phi(X_J) \neq 0$ is finite and let Δ_Φ denote the union of all such subsets.

9.2   Thermodynamical Limit, Existence of Free Energy

Denote by $\mathfrak{B}_1$ the class of potentials with the norm

$$\|\Phi\| = \sum_{0\in J\in W} |J|^{-1} \sup_{x\in X_J} |\Phi(x)| < \infty$$

and let for $I \in W$, $\bar{x} \in X_{\mathbb{Z}^\nu\setminus I}$,

$$f_I^{\overline{x}}(\Phi) = \frac{\ln Z_I(\overline{x}, \Phi)}{|I|} \ . \tag{9.2.1}$$

__Theorem 9.2.1.__ Suppose $\Phi \in \mathcal{B}_1$. Then for any sequence of subsets $I_n \in W$, $n \in \mathbb{N}$ tending to infinity in the sense of Van Hove and for any sequence of boundary conditions $\overline{x}_n \in X_{\mathbb{Z}^\nu \setminus I_n}$, $n \in \mathbb{N}$ the following limit exists:

$$\lim_{n \to \infty} f_{I_n}^{\overline{x}_n}(\Phi) = f(\Phi) . \tag{9.2.2}$$

The function $f(\Phi)$ is called the free energy.

__Proof.__ Since the space of finite potentials $\mathcal{B}_0$ is everywhere dense in $\mathcal{B}_1$ then it is sufficient to prove Theorem 9.2.1 in the case of finite potential. Let's prove some preliminary statements.

__Lemma 9.2.1.__ If $\Phi, \Psi \in \mathcal{B}_1$ then

$$|f_I^{\overline{x}}(\Phi) - f_I^{\overline{x}}(\Psi)| \leq \|\Phi - \Psi\|, \quad \overline{x} \in X_{\mathbb{Z}^\nu \setminus I}. \tag{9.2.3}$$

__Proof.__ We have

$$|f_I^{\overline{x}}(\Phi) - f_I^{\overline{x}}(\Psi)| = \frac{1}{|I|} \left| \ln \frac{Z_I(\overline{x}, \Phi)}{Z_I(\overline{x}, \Psi)} \right| =$$

$$= \frac{1}{|I|} \left| \ln \left(1 + \frac{\int_{X_I} \exp\{-U_I^{\overline{x},\Psi}(x)\} \left(\exp\{-U_I^{\overline{x},\Phi}(x) + U_I^{\overline{x},\Psi}(x)\} - 1 \right) \mu_I(dx)}{\int_{X_I} \exp\{-U_I^{\overline{x}}(x)\} \mu_I(dx)} \right) \right| \leq$$

$$\leq \frac{1}{|I|} \sup_{x \in X_I, \ \overline{x} \in X_{\mathbb{Z}^\nu \setminus I}} |U_I^{\overline{x}, \Phi - \Psi}(x)| \leq \|\Phi - \Psi\|.$$

<u>Corollary 9.2.1.</u> If $\Phi \in \mathcal{B}_1$ then

$$\left| f_I^{\bar{x}}(\Phi) - \ln \mu(X_I) \right| \le \|\Phi\|.$$

For the proof it is sufficient to put $\psi \equiv 0$ in (9.2.3).

<u>Lemma 9.2.2.</u> Let $\Phi \in \mathcal{B}_0$ and for any $V_1, V_2 \in W$, $V_1 \cap V_2 = \emptyset$ $N(V_1, V_2)$ denotes the number of points $a \in \mathbb{Z}^\nu$ such that

$$V_1 \cap (\Delta_\Phi + a) \neq \emptyset, \qquad V_2 \cap (\Delta_\Phi + a) \neq \emptyset.^{[*]}$$

Then

$$\left| \ln Z_{V_1 \cup V_2}(\Phi) - \ln Z_{V_1 \cup V_2}(\Phi) - \ln Z_{V_2}(\Phi) \right| \le N(V_1, V_2) \|\Phi\|,$$

where

$$Z_I(\Phi) = \int_{X_I} \exp\{-U_I(x)\} \mu_I(dx), \qquad U_I(x) = \sum_{J \subset I} \Phi(x_J).$$

<u>Proof.</u> We have

$$\left| \ln \frac{Z_{V_1 \cup V_2}(\Phi)}{Z_{V_1}(\Phi) Z_{V_2}(\Phi)} \right| = \left| \ln \left(1 + \frac{Z_{V_1 \cup V_2}(\Phi) - Z_{V_1}(\Phi) Z_{V_2}(\Phi)}{Z_{V_1}(\Phi) Z_{V_2}(\Phi)} \right) \right|.$$

Further

$$\left| \int_{X_{V_1 \cup V_2}} \exp\{-U_{V_1 \cup V_2}(x)\} \mu_{V_1 \cup V_2}(dx) - \int_{X_{V_1}} \exp\{-U_{V_1}(x)\} \mu_{V_1}(dx) \times \right.$$

$$\left. \times \int_{X_{V_2}} \exp\{-U_{V_2}(x)\} \mu_{V_2}(dx) \right| \le \int_{X_{V_1 \cup V_2}} \exp\{-U_{V_1}(x) - U_{V_2}(y)\} \times$$

$$\times \left| \exp\{-U_{V_1 \cup V_2}(xy) + U_{V_1}(x) + U_{V_2}(y)\} - 1 \right| \mu_{V_1 \cup V_2}(dx\,dy) \le$$

$$\le \sup_{(x,y) \in X_{V_1 \cup V_2}} \left| \exp\{-U_{V_1 \cup V_2}(xy) + U_{V_1}(x) + U_{V_2}(y)\} - 1 \right| Z_{V_1}(\Phi) Z_{V_2}(\Phi) \le$$

$[*]$ $\Delta_\Phi = \{J : \Phi(J) \neq \phi\}$

$$\leq \left\{ \exp\left\{ \sum_{\substack{J \subset V_1 \cup V_2 \\ J \cap V_1 \neq \varnothing,\ J \cap V_2 \neq \varnothing}} \sup_{x \in X_J} |\Phi(x)| \right\} - 1 \right\} Z_{V_1}(\Phi) Z_{V_2}(\Phi).$$

By virtue of the finiteness of potential Φ, $\Phi(x) \neq 0$, $x \in X_I$, iff there exists a point $a \in \mathbb{Z}^\nu$ such that $a + I \subset \Delta_\Phi$, i.e., iff

$$V_1 \cap (\Delta_\Phi + a) \neq \varnothing, \quad V_2 \cap (\Delta_\Phi + a) \neq \varnothing.$$

Hence

$$\left| \ln \frac{Z_{V_1 \cup V_2}(\Phi)}{Z_{V_1}(\Phi) Z_{V_2}(\Phi)} \right| \leq \sum_{\substack{J \subset V_1 \cup V_2 \\ J \cap V_1 \neq \varnothing,\ J \cap V_2 \neq \varnothing}} \sup_{x \in X_J} |\Phi(x)| \leq N(V_1 \cup V_2)\|\Phi\|.$$

Lemma 9.2.2 is proven.

Now let's prove (9.2.2) . We will use the notations of Section 7.1. Let $\Gamma_s^- \subset I$ be the union of all subsets Λ_t such that $\Lambda_t \subset I$ (of course, $|\Gamma_s^-| = N_s^-(I)$) and let these cubes be somehow enumerated $\Lambda^{(j)}$, $j = 1,2,\ldots, N_s^-(I)$. Denote

$$\Gamma(j) = (I \setminus \Gamma_s^-) \cup \bigcup_{i=1}^{j} \Lambda^{(i)}, \quad j = 1,2,\ldots, N_s^-(I).$$

Hence

$$\Gamma(N_s^-(I)) = I.$$

According to Lemma 9.2.2 we can write

$$\left| \ln Z_{\Gamma(j)}(\Phi) - \ln Z_{\Gamma(j-1)}(\Phi) - \ln Z_{\Lambda^{(j)}}(\Phi) \right| \leq$$

$$\leq N(\Gamma(j-1), \Lambda^{(j)})\|\Phi\|, \quad j = 1,2,\ldots, N_s^-(I).$$

Summing over j **we obtain**

$$\left| \ln Z_I(\Phi) - \ln Z_{I \setminus \Gamma_s^-}(\Phi) - N_s^-(I) \ln Z_{\Lambda(s)}(\Phi) \right| \leq$$

$$\leq \|\Phi\| \sum_{j=1}^{N_s^-(I)} N(\Gamma(j-1), \Lambda^{(j)}).$$

By the definition $N(\Gamma(j-1), \Lambda^{(j)})$ is less than the number of points β such that $\Delta_\Phi + \beta$ intersects $\Lambda^{(j)}$ and its complement simultaneously. Therefore for any $\varepsilon > 0$ one can choose such s that

$$|\Lambda(s)|^{-1} N(\Gamma(j-1), \Lambda^{(j)}) < \frac{\varepsilon}{2\|\Phi\|}$$

whence

$$|I|\,(f_I(\Phi) - f_{\Lambda(s)}(\Phi)) - |I \setminus \Gamma_s^-|\,(f_{I \setminus \Gamma_s^-}(\Phi) -$$

$$- f_{\Lambda(s)}(\Phi)| \leq N_s^-(I)\,|\Lambda(s)|\,\frac{\varepsilon}{2}\,.$$

Using Corollary 9.2.1 we find

$$|f_{I \setminus \Gamma_s^-}(\Phi) - f_{\Lambda(s)}(\Phi)| \leq 2\|\Phi\|\,.$$

On the other hand since $I \to \infty$ in a sense of Van Hove then

$$\frac{|I \setminus \Gamma_s^-|}{|I|} \leq \frac{N_s^+(I) - N_s^-(I)}{N_s^+(I)} < \frac{\varepsilon}{4\|\Phi\|}\,.$$

From this last two relations we obtain

$$|f_I(\Phi) - f_{\Lambda(s)}| < \varepsilon$$

which proves the theorem in the case of empty boundary conditions.

Let's consider the general case. Denote for $I \in W$

$$(I)_d = \{t \in \mathbb{Z}^\nu : \tau(t, I) \leq d\}, \quad d \in R_+^1$$

$$\tau(t, I) = \min_{s \in I} \{\tau(t, s)\}, \quad \tau(t, s) = \max_{1 \leq i \leq \nu} |t^{(i)} - s^{(i)}|\,.$$

We have

$$|f_I(\Phi) - f_I^{\tilde{x}}(\Phi)| = \frac{1}{|I|}\,|\ell n\,(1 + \frac{Z_I^{\tilde{x}}(\Phi) - Z_I(\Phi)}{Z_I(\Phi)})| \leq$$

$$\leq \frac{1}{|I|} \sum_{J \in W(I)} \sum_{\varnothing \neq \tilde{J} \in W(\mathbb{Z}^\nu \setminus I)} \sup_{x \in X_J,\, \tilde{x} \in X_{\tilde{J}}} |\Phi(x\tilde{x})| \leq$$

$$\leq \frac{1}{|I|} \sum_{\substack{J \cap I \neq \varnothing \\ J \subset (I)_{d_0}}} \sup_{x \in X_J} |\Phi(x)|,$$

where d_0 is the range of interaction of finite potential Φ (i.e. $\Phi(x_J) = 0$ if $\max\limits_{t,s \in J} \{ r(t,s) \} > d_0$, $J \in W$).

Whence

$$| f_I(\Phi) - f_I^{\bar{x}}(\Phi) | \leq \frac{|(I)_c|}{|I|} \, \|\Phi\| \to 0, \quad |I| \to \infty.$$

Theorem 9.2.1 is proven.

9.3 Strong Convexity of Free Energy, Linear Growth of Variance of Energy

Let's formulate several preliminary definitions. Suppose E is some Banach space and $h(x)$ is some function defined on E. It is said that $h(x)$ is convex on set S in the directions belonging to cone $K \subseteq E$ if for any $x, x+e \in S$, $e \in K$ and for any $\lambda \in [0,1]$

$$\Delta^{\lambda}_{x,e} h \equiv \lambda h(x+e) + (1-\lambda) h(x) - h(x+\lambda e) \geq 0. \tag{9.3.1}$$

If in (9.3.1) the strict inequality is valid then the function $h(x)$ is called strict convex on set S in the directions belonging to cone K. In this terms the usual definition of the convex (strict convex) function on some convex set M is equivalent to the following : function $h(x)$ is convex (strict convex) on set M in the directions of cone E. We say that function $h(x)$ is strong convex on set S in the directions of cone K if for each N, $0 < N < \infty$, there exists a constant $C_N > 0$ such that for any $x, x+e \in S$, $e \in K$, $\|e\| \leq N$ and for any $\lambda \in [0,1]$ the following relation is valid:

$$\Delta^{\lambda}_{x,e} h \geq \lambda (1-\lambda) C_N \|e\|^2. \tag{9.3.2}$$

Note, if $x \in S$ and $e \in K$, $\|e\| \leq N$ are such that $x + \lambda e \in S$ for any $\lambda \in [\alpha, \beta] \subset [0,1]$ and if function $h(x+\lambda e)$ is twice differentiable with respect to λ, $\lambda \in [\alpha, \beta]$, then from (9.3.2) it follows that

$$\frac{\partial^2 h(x+\lambda e)}{\partial \lambda^2} \geqslant C_N \, \|e\|^2, \qquad \lambda \in [\alpha, \beta].$$

We say that potential $\Phi \in \mathcal{B}$ is degenerate if there exist a function $F_{\{0\}}^{\Phi}(\bar{x})$, $\bar{x} \in X_{\mathbb{Z}^\nu \setminus \{0\}}$ and a set $X' \subset X$, $\mu(X') = \mu(X)$ such that

$$\mathcal{U}_{\{0\}}^{\bar{x}, \Phi}(x) \equiv F_{\{0\}}^{\Phi}(\bar{x}), \qquad x \in X'. \tag{9.3.3}$$

The existence of some function $F_I^{\Phi}(\bar{x})$, $\bar{x} \in X_{\mathbb{Z}^\nu \setminus I}$, $I \in W$ such that

$$\mathcal{U}_I^{\bar{x}, \Phi}(x) = F_I^{\Phi}(\bar{x}) \qquad \text{for any } x \in X_I', \ \bar{x} \in X_{\mathbb{Z}^\nu \setminus I}$$

follows from (9.3.3).

Really, for $t \in I$ we have

$$\mathcal{U}_I^{\bar{x}}(x) = \sum_{J: t \in J \in W} \Phi(x_J) + \sum_{J: t \notin J \in W} \Phi(x_J). \tag{9.3.4}$$

Second sum in (9.3.4) does not depend on t. First sum also does not depend on t according to (9.3.1). Since t is arbitrary we have (9.3.2). Note that the Gibbs random fields corresponding to the potentials Φ and $\Phi + \mathcal{U}$ are identical if $\Phi, \mathcal{U} \in \mathcal{B}$ and $\mathcal{U}$ is a degenerate potential. Note also that any vacuum potential represents an example of non-degenerate potential.

The main result of this section is the following theorem.

Theorem 9.3.1(/34/). Let $S \subset \mathcal{B}$ be some compact set and $K \subseteq \mathcal{B}$ be a cone. Let $K \cap E^{(1)}$ be compact, where $E^{(1)}$ is the sphere of radius 1 in $\mathcal{B}$. If the set $K \cap E^{(1)}$ does not contain degenerate potentials then the function $f(\Phi)$ is strong convex on S in the directions of cone K.

Proof. Let's note some general probabilistic facts. Let $(\Omega, \mathcal{F}, P)$ be a measurable space with a measure μ such that

$$\Omega = \Omega_1 \times \Omega_2, \qquad \mathcal{F} = \mathcal{F}_1 \times \mathcal{F}_2, \qquad \mu = \mu_1 \times \mu_2.$$

If $p(\omega) = p(\omega_1, \omega_2)$, $\omega_1 \in \Omega_1$, $\omega_2 \in \Omega_2$, is some density of probability distribution on Ω , then the conditional density is defined in the following way

$$p\left(\omega_2/\omega_1\right) = p\left(\omega_2, \omega_1\right)/\bar{p}\left(\omega_1\right),$$

$$\bar{p}\left(\omega_1\right) = \int_{\Omega_2} p\left(\omega_2, \omega_1\right)\mu_2\left(d\omega_2\right).$$

Conditional expectation and conditional variance of a random variable $\xi(\omega)$ are defined with the help of the relations

$$E\left(\xi/\omega_1\right) = \int_{\Omega_2} \xi(\omega_1,\omega_2)\, p\left(\omega_2/\omega_1\right)\mu_2\left(d\omega_2\right)$$

and

$$D\left(\xi/\omega_1\right) = \int_{\Omega_2} \left(\xi(\omega_1,\omega_2)-E\left(\xi/\omega_1\right)\right)^2 p\left(\omega_2/\omega_1\right)\mu_2\left(d\omega_2\right),$$

respectively.

It is easy to verify that

$$D\xi = \int_{\Omega} \left(\xi-E\xi\right)^2 p(\omega)\mu\left(d\omega\right) = \int_{\Omega_1} D\left(\xi/\omega_1\right)\bar{p}\left(\omega_1\right)\mu_1\left(d\omega_1\right) +$$

$$+ \int_{\Omega_1} \left(E\left(\xi/\omega_1\right)-E\xi\right)^2 \bar{p}\left(\omega_1\right)\mu_1\left(d\omega_1\right)$$

and whence

$$D\xi \geqslant \int_{\Omega_1} D\left(\xi/\omega_1\right)\bar{p}\left(\omega_1\right)\mu\left(d\omega_1\right). \qquad (9.3.5)$$

Further by $E_{I,\Phi}^{\bar{x}}$, $D_{I,\Phi}^{\bar{x}}$ and $Cov_{I,\Phi}^{\bar{x}}$ we will denote the expectation, variance and covariance with respect to the Gibbs distribution $q_I^{\Phi,\bar{x}}(x)$, $x \in X_I$, $x \in X_{\mathbb{Z}^{\nu}\setminus I}$ correspondingly.

Now let's prove the following important lemma.

<u>Lemma 9.3.1.</u> Suppose that the conditions of Theorem 9.3.1 are fulfilled. Then for each N, $0 < N < \infty$ there exists a constant $C_N > 0$ and an increasing sequence of Cubes I_n, $n=1,2...,$ $\underset{n}{\cup} I_n = \mathbb{Z}^{\nu}$ such that for any $\Phi_0 \in S$, $\Phi \in K \cap E^{(N)}$ ($E^{(N)}$ is the sphere of radius N in $\mathcal{B}$) the following inequality is valid:

$$D_{I_n,\Phi_\lambda}^{\bar{x}}(\mathcal{U}_{I_n}^{\bar{x}_n}) \geq 2C_N |I_n| \,\|\Phi\|^2, \qquad\qquad (9.3.6)$$

$$\Phi_\lambda = \Phi_0 + \lambda\Phi, \quad \lambda \in [0,1].$$

Note that Lemma 9.3.1 implies the statement of Theorem 9.3.1.
Really, we have

$$\frac{\partial^2 f_{I_n}^{\bar{x}_n}(\Phi_\lambda)}{\partial\lambda^2} = \frac{D_{I_n,\Phi_\lambda}^{\bar{x}_n}(\mathcal{U}_I^{\bar{x}_n})}{|I_n|}, \qquad\qquad (9.3.7)$$

it follows from (9.3.6) and (9.3.5) that

$$\frac{\partial^2 f_{I_n}^{\bar{x}_n}(\Phi)}{\partial\lambda^2} \geq 2C_N \|\Phi\|^2$$

for any $\lambda \in [0,1]$ and hence

$$\Delta_{\Phi^0,\Phi}^\lambda f_{I_n}^{\bar{x}_n} \geq \lambda(1-\lambda) C_N \|\Phi\|^2$$

for any $\Phi \in K \cap E^{(N)}$, $\Phi_0 \in S$.
The latter proves Theorem 9.3.1.

<u>Proof of Lemma 9.3.1.</u> Let $J, V \subset \mathbb{Z}^\nu$, $J \cap V = \varnothing$, $|J| < \infty$.
Define the function $\Phi_J^{\bar{x}}(x)$, $x \in X_J$, $\bar{x} \in X_V$ putting

$$\Phi_J^{\bar{x}}(x) = \sum_{\bar{J}:\bar{J}\in W(V)} \Phi(x\bar{x}_{\bar{J}}), \quad x \in X_J, \quad \Phi_\varnothing^{\bar{x}} = 0.$$

Then for $V = \mathbb{Z}^\nu \setminus I$, $|I| < \infty$, we can write

$$\mathcal{U}_I^{\bar{x}}(x) = \sum_{J \subset I} \Phi_J^{\bar{x}}(x_J), \quad x \in X_I, \quad \bar{x} \in X_{\mathbb{Z}^\nu\setminus I} \qquad (9.3.8)$$

and it is clear that

$$\mathcal{U}_{\{s\}}^{\bar{x}}(x) = \Phi_{\{s\}}^{\bar{x}}(x), \quad s \in \mathbb{Z}^\nu, \quad x \in X_s, \quad \bar{x} \in X_{\mathbb{Z}^\nu\setminus\{s\}}.$$

Using inequality (9.3.5) we have

$$D_{I,\Phi_\lambda}^{\bar{x}}(\mathcal{U}_I^{\bar{x}}) \geq \int_{X_{I\setminus I_i}} D_{I,\Phi_\lambda}^{\bar{x}}(\mathcal{U}_I^{\bar{x}}/y)(q_I^{\bar{x}})_{I_i}(y)\,\mu_{I_i}(dy) =$$

$$= \int_{X_{I \smallsetminus I_\iota}} D^{\bar{x}y}_{I_\iota, \Phi_\lambda} (\, U^{\bar{x}y}_{I_\iota})(q^{\bar{x}}_I)_{I_\iota}(y)\, \mu_{I_\iota}(dy),$$

where

$$(q^{\bar{x}}_I)_{I_\iota}(y) = \int_{X_{I \smallsetminus I_\iota}} q^{\bar{x}}_I (yz)\, \mu_{I \smallsetminus I_\iota}(dz), \quad y \in X_{I_\iota}.$$

For convenience let's denote

$$D^{\bar{x}}_{I_\iota, \Phi_\lambda} = \bar{D}, \quad Cov^{\bar{x}}_{I_\iota, \Phi_\lambda} = \overline{Cov}, \quad x' = (y\bar{x}) \in X_{\mathbb{Z}^\nu \smallsetminus I_\iota}.$$

We have

$$\bar{D}(U^{x'}_I) = \bar{D}\Big(\sum_{s \in I_\iota} \Phi^{x'}_{\{s\}} + \sum_{J: J \subset I_\iota,\, |J| \geqslant 2} \Phi^{x'}_J \Big) =$$

$$= \sum_{s \in I_\iota} \Big(\bar{D}\Phi^{x'}_{\{s\}} + \overline{Cov}\,\big(\Phi^{x'}_{\{s\}}, \sum_{J: J \subset I_\iota,\, |J| \geqslant 2} \Phi^{x'}_J \big) +$$

$$+ \overline{Cov}\,\big(\Phi^{x'}_{\{s\}}, \sum_{J: J \subset I_\iota,\, J \neq \{s\}} \Phi^{x'}_J \big)\Big) + \bar{D}\Big(\sum_{J: J \subset I_\iota} \Phi^{x'}_J \Big) \geqslant$$

$$\geqslant \sum_{s \in I_\iota} \big[\, \bar{D}\Phi^{x'}_{\{s\}} + \overline{Cov}\,\big(\Phi^{x'}_{\{s\}}, \sum_{J: J \subset I_\iota,\, J \neq \{s\}} \Phi^{x'}_J \big) +$$

$$+ \overline{Cov}\,\big(\Phi^{x'}_{\{s\}}, \sum_{J: J \subset I_\iota,\, |J| \geqslant 2} \Phi^{x'}_J \big)\big].$$

Note that

$$\overline{Cov}\,\big(\Phi^{x'}_{\{s\}} \sum_{J: J \subset I_\iota,\, J \neq \{s\}} \Phi^{x'}_J \big) = \overline{Cov}\,\big(\Phi^{x'}_{\{s\}}, \sum_{J: s \notin J \subset I_\iota} \Phi^{x'}_J \big) +$$

$$+ \overline{Cov}\,\big(\Phi^{x'}_{\{s\}}, \sum_{J: s \in J \subset I_\iota,\, |J| \geqslant 2} \Phi^{x'}_J \big).$$

Fix some $s \in I_l$ and denote $B_{I_l}^{\{s\}} = \{ J : J \subset I_l,\ s \notin J \}$.

Let's consider the partition of set $B_{I_l}^{\{s\}}$ which is defined in the following way

$$B_{I_l}^{\{s\}} = \bigcup_{k \in I_l \setminus \{s\}} T_{k,s} \ ,$$

where the set $J \in B_l^{\{s\}}$ belongs to $T_{k,s}$ if the next conditions are valid.

a) $k \in J$;

b) $\varphi(k,s) = \max_{e \in J} \{ \overline{\varphi}(\{\ell\}, \{k\}) \}.$

Here $\overline{\varphi}(J,V)$, $J, V \in W$ is the φ -mixing coefficient between σ -fields $\mathcal{F}_J$ and $\mathcal{F}_V$ computed with respect to Gibbs random field $q_I^{\bar{x}}(x)$. If in condition b)   max is achieved on more then one point we include the set J in arbitrary $T_{k,s}$.

Further we can write

$$\left| \overline{\mathrm{Cov}}\left(\Phi_{\{s\}}^{x'}, \Phi_J^{x'} \right) \right| \le \sup_{x \in X_s} \left| \Phi_{\{s\}}^{x'}(x) \right| \sup_{x \in X_J} \left| \Phi_J^{x'}(x) \right| \overline{\varphi}(\{s\}, J) \le$$

$$\tag{9.3.9}$$

$$\le \sup_{x \in X_s} \left| \Phi_{\{s\}}^{x'}(x) \right| \sup_{x \in X_J} \left| \Phi_J^{x'}(x) \right| |J| \, \overline{\varphi}(k,s)$$

and

$$\left| \overline{\mathrm{Cov}}\left(\Phi_{\{s\}}^{x'}, \sum_{J \in B_{I_l}^s} \Phi_J^{x'} \right) \right| \le \sum_{k \in I_l \setminus \{s\}} \sum_{J \in T_{k,s}} \sup_{x \in X_s} \left| \Phi_{\{s\}}^{x'}(x) \right| \times$$

$$\times \sup_{x \in X_J} \left| \Phi_J^{x'}(x) \right| |J| \, \overline{\varphi}(k,s) \le \| \Phi \| \sum_{k \in I_l \setminus \{s\}} \overline{\varphi}(s,k) \times$$

$$\tag{9.3.10}$$

$$\sum_{J \in T_{k,s}} |J| \sup_{x \in X_J} \left| \Phi_J^{x'}(x) \right| \le \| \Phi \|^2 \sum_{k \in I_l \setminus \{s\}} \overline{\varphi}(s,k) \le$$

$$\le \sum_{k \in \mathbb{Z}^\nu \setminus \{s\}} \left(\rho_{ks} + \sum_{\ell \in \mathbb{Z}^\nu \setminus \{s\}} \rho_{ke} \rho_{es} + \sum_{e,n \in \mathbb{Z}^\nu \setminus \{s\}} \rho_{ke} \rho_{en} \rho_{ns} + \cdots \right).$$

Denote

$$\mathbb{Z}_d = \{ td : t \in \mathbb{Z}^\nu \} \subseteq \mathbb{Z}^\nu, \quad d \in \mathbb{Z}^\nu_+,$$

and associate to any $s \in \mathbb{Z}^\nu$ the ν-dimensional cube $\beta_s^d \subset \mathbb{Z}^\nu$ defined in the following way:

$$\beta_s^d = \{ t \in \mathbb{Z}^\nu : s^{(i)} - \frac{d}{2} < t^{(i)} \le s^{(i)} + \frac{d}{2}, \quad i = \overline{1,\nu} \}.$$

Now let I_1 $(|I_1| \ge 2)$ be a cube from the sublattice $\mathbb{Z}^\nu_d$. Put

$$I = \bigcup_{s \in I_1} \beta_s^d, \qquad R_\tau^S = \sup_{\Phi \in S} \{ R_\tau^\Phi \},$$

$$R_\tau^\Phi = \sum_{\substack{J : 0 \in J \in W \\ D(J) \ge \tau}} |J| \sup_{x \in X_J} |\Phi(x)|,$$

where $D(J)$ is the diameter of J.
Using the continuity and monotonicity of function R_τ^Φ with respect to Φ and τ correspondingly, one can obtain the next lemma.

<u>Lemma 9.3.2.</u> For any compact $S \subset \mathcal{B}$ and any increasing sequence I_n, $n = 1, 2, \ldots$ such that $\bigcup_n I_n = \mathbb{Z}^\nu$

$$\lim_{n \to \infty} R_{D(I_n)}^S = 0.$$

Since

$$\sup_{x \in X_S, \, z \in X_K, \, \bar{y} \in X_{I_1 \setminus \{s,\kappa\}}} | \Phi_{\lambda, \{s,\kappa\}}^{x'\bar{y}}(x,z) | \le R_d^S + R_d^{S_N},$$

then for sufficiently large d using the following inequality $|e^{-x} - 1| < 2|x|$, $0 < x < \frac{1}{2}$ we can write

$$\rho_{st} \le \frac{1}{2} e^{4 \|\Phi\|} \sup_{x \in X_S, \, z \in X_K, \, \bar{y} \in X_{I_1 \setminus \{s,\kappa\}}} | \Phi_{\lambda \{s,\kappa\}}^{x'y}(xz) |.$$

From the compactness of the sets S and S_N it follows the existence of a constant $C_N^{(1)}$ such that for $\Phi^0 \in S$ and $\Phi \in S_N$

$$e^{4 \|\Phi_\lambda\|} \le C_N^{(1)}.$$

Thus for sufficiently large d we obtain

$$\sum_{k \in I_1 \setminus \{s\}} \rho_{ks} \leq C_N^{(1)} \sum_{k \in \mathbb{Z}^\nu \setminus \{s\}} \sup_{\substack{x \in X_s,\, \bar{y} \in X_{I_1} \setminus \{s,k\} \\ z \in X_k}} \left| \Phi_{\lambda,\{s,k\}}^{x'y}(xz) \right| \leq$$

$$\leq C_N^{(1)} \sum_{k \in \mathbb{Z}^\nu \setminus \{s\}} \sum_{J:s,k \in J \in W} \sup_{x \in X_J} \left| \Phi(x) \right| \leq$$

$$\leq C_N^{(1)} \sum_{\substack{J:s,k \in J \in W \\ s \neq k}} |J| \sup_{x \in X_J} \left| \Phi_\lambda(x) \right| \leq C_N^{(1)} \left(R_d^S + R_d^{S_n} \right).$$

Since for large d $\quad C_N^{(1)}\left(R_d^S + R_d^{S_N} \right) < \frac{2}{3}$ then from $(9.3.10)$ we deduce

$$\sum_{k \in I_1 \setminus \{s\}} \bar{\varphi}(s,k) \leq 2\, C_N^{(1)} \left(R_d^S + R_d^{S_N} \right). \qquad (9.3.11)$$

Substituting $(9.3.11)$ into $(9.3.9)$ and using the boundedness of $\| \Phi \|$ on S_N we finally obtain

$$\left| \overline{\mathrm{Cov}}\left(\Phi_{\{s\}}^{x'}, \sum_{J \in B_{I_1}^S} \Phi_J^{x'} \right) \right| \leq C_N^{(2)} \left(R_d^S + R_d^{S_N} \right). \qquad (9.3.12)$$

Further

$$\left| \overline{\mathrm{Cov}}\left(\Phi_{\{s\}}^{x'}, \sum_{\substack{J:\, s \in J \subset I_1 \\ |J| \geq 2}} \Phi_J^{x'} \right) \right| \leq$$

$$\leq \sup_{x \in X_s} \left| \Phi_{\{s\}}^{x'}(x) \right| \sup_{x \in X_{I_1}} \left| \sum_{\substack{J:\, s \in J \subset I_1 \\ |J| \geq 2}} \Phi_J^{\bar{x}}(x_I) \right| \leq$$

$$\leq \| \Phi \|^2 R_d^{S_N} \leq C_N^{(3)} R_d^{S_N}.$$

Thus for sufficiently large d

$$\left| \overline{Cov}\left(\Phi_{\{s\}}^{x'}, \sum_{J:J\subset I_L, |J|\neq\{s\}} \Phi_J^{x'}\right)\right| \leq C_N^{(2)}\left(R_d^s + R_d^{s_N}\right) + C_N^{(3)} R_d^{s_N}. \qquad (9.3.13)$$

Similarly one can obtain the following inequality

$$\left| \overline{Cov}\left(\Phi_{\{s\}}^{x'}, \sum_{J:J\subset I_L, |J|\geqslant 2} \Phi_J^{x'}\right)\right| \leq C_N^{(2)}\left(R_d^s + R_d^{s_N}\right) + C_N^{(3)} R_d^{s_N}. \qquad (9.3.14)$$

Further we have for $\quad x' = (x^0\, \bar{x}) \in X_{\mathbb{Z}^\nu \setminus I_L}$,

$$y = x^0_{\beta_s^d \setminus \{s\}} \in X_{\beta_s^d \setminus \{s\}},$$

$$\overline{D}\left(\Phi_{\{s\}}^{x'}\right) = \overline{D}\left(\Phi_{(s)}^{y} + \left(\Phi_{\{s\}}^{x'} - \Phi_{\{s\}}^{y}\right)\right) \geqslant$$

$$\geqslant \overline{D}\left(\Phi_{\{s\}}^{y}\right) + 2\,\overline{Cov}\left(\Phi_{\{s\}}^{y}, \Phi_{\{s\}}^{x'} - \Phi_{\{s\}}^{y}\right)$$

and

$$\overline{Cov}\left(\Phi_{\{s\}}^{y}, \Phi_{\{s\}}^{x'} - \Phi_{\{s\}}^{y}\right) \leqslant \qquad\qquad (9.3.15)$$

$$\leqslant 2 \sup_{x\in X_s}\left|\Phi_{\{s\}}(x)\right| \sup_{x\in X_{\{s\}}}\left|\Phi_{\{s\}}^{x'}(x) - \Phi_{\{s\}}^{y}(x)\right| \leqslant C_N^{(3)} R_d^{s_N}.$$

Let

$$C(\Phi) = \sup_{x'\in X_{\mathbb{Z}^\nu \setminus \{s\}}} \sup_{x, z \in X_s} \left| \mathcal{U}_{\{s\}}^{x',\Phi}(x) - \mathcal{U}_{\{s\}}^{x',\Phi}(z)\right|, \quad \Phi\in\mathcal{B}.$$

Note, that the function $C(\Phi)$ is continuous. Since the set S_N is compact and does not contain degenerate potentials then

$$\inf_{\Phi\in S_N} C(\Phi) \geqslant C_N^{(4)} > 0. \qquad (9.3.16)$$

From (9.3.16) and Lemma 9.3.2 it follows that for sufficiently large d

$$\inf_{\Phi\in S_N}\ \sup_{y\in X_{\beta_s^d\setminus\{s\}}}\ \sup_{x,z\in X}\ \left|\mathcal{U}_{\{s\}}^{x',\Phi}(x)-\mathcal{U}_{\{s\}}^{x',\Phi}(z)\right|\geq C_N^{(5)}>0.$$

The latter implies for each $\Phi\in S_N$ the existence of the set $A_\Phi\subset X_s'$, $\mu(A_\Phi)>0$ and boundary conditions $y_\Phi\in X_{\beta_s^d\setminus\{s\}}$ such that

$$\sup_{x,z\in A_\Phi}\left|\mathcal{U}_{\{s\}}^{y_\Phi}(x)-\mathcal{U}_{\{s\}}^{y_\Phi}(z)\right|\geq C_N^{(6)}>0. \qquad (9.3.17)$$

Then we have

$$\bar{D}(\mathcal{U}_{\{s\}}^{y_\Phi})\geq E\bar{D}(\Phi_{\{s\}}^{y_\Phi}/A_\Phi)\geq$$

$$\geq(C_N^{(6)})^2\int_X\int_{A_\Phi\setminus\{z\}}q_{\{s\}}^{y_\Phi}(x)\mu_s(dx)\mu(dz)\times$$

$$\times\left(\int_{A_\Phi}q_{\{s\}}^{y_\Phi}(x)\mu_s(dx)\right)^{-2}\geq$$

$$\geq(C_N^{(6)})e^{4\|\Phi\|}\mu^{-2}(A_\Phi)\int_X\mu^2(A_\Phi\setminus\{z\})\mu(dz). \qquad (9.3.18)$$

From (9.3.18) and the compactness of sets S and S_N we deduce

$$\bar{D}(\mathcal{U}_{\{s\}}^{y_\Phi})\geq C_N^{(7)}>0. \qquad (9.3.19)$$

Further using (9.3.19), (9.3.11), (9.3.12), (9.3.13), (9.3.14) and lemma 9.3.2 we obtain

$$\bar{D}(\mathcal{U}_{I_i}^{\bar{x}})\geq C_N^{(8)}>0 \qquad (9.3.20)$$

for any $\bar{x}$ such that $\bar{x}=(x^0,x')$, $x_{\beta_s^d\setminus\{s\}}^0=y_\Phi$ and sufficiently large d .

The latter implies

$$\bar{D}\left(\mathcal{U}_I^{\bar{x}}/x_o\right) \geq C_N^{(9)}\,|I_i|, \qquad C_N^{(9)} > 0$$

and

$$D_{I,\Phi_\lambda}^{\bar{x}}\left(\mathcal{U}_I^{\bar{x}}\right) \geq C_N^{(9)}\,|I_i| \int_{A_\Phi} \left(q_{I,\Phi_\lambda}^{\bar{x}}\right)_\beta (x)\,\mu_\beta\,(dx) \geq$$

$$\geq C_N^{(10)}\,|\beta|\,|I_i|\,\|\Phi\|\,\mu\,(A_\Phi), \qquad \beta = \beta_s^d \setminus \{s\}.$$

Finally we have

$$D_{I,\Phi_\lambda}^{\bar{x}}\left(\mathcal{U}_I^{\bar{x},\Phi}\right) \geq 2 C_N \|\Phi\|^2, \qquad \Phi \in S_N,$$

$$\tag{9.3.21}$$

$$C_N = C_N^{(9)} \inf_{\Phi \in S} \{\|\Phi\|\}.$$

Now let $\widetilde{I}_n \subset \mathbb{Z}^\nu d$ (d is sufficiently large) be some increasing sequence of ν -dimensional cubes. Putting

$$I_n = \sum_{s \in \widetilde{I}_n} \beta_s^d \subset \mathbb{Z}^\nu, \qquad n \in \mathbb{N},$$

we obtain the required sequence of cubes which satisfy (9.3.21) for any $n \in \mathbb{N}$. This completes the proof of Lemma 9.3.1. Other results concerning these questions can be found in /53/,/40/.

9.4 Limit Theorems for Gibbs Random Fields

The aim of this section is to prove the following theorems.

<u>Theorem 9.4.1.</u> Let ξ_t, $t \in \mathbb{Z}^\nu$, be a centered Gibbs random field taking the values in some compact metric space X . Let the potential $\Phi \in \mathcal{B}$ corresponding to this Gibbs random field be translation invariant and such that

$$\frac{1}{2} e^{4\|\Phi\|}\left(e^{4\|\Phi\|} - 1\right) < 1, \tag{9.4.1}$$

$$\sum_{J:\, 0 \in J \in W} |J|\,(D(J))^\gamma \sup_{x \in X_J} |\Phi(x)| < \infty, \qquad \gamma > 2, \tag{9.4.2}$$

$D(J)$ is the diameter of d .

Let $q(x)$, $x \in X$ be a measurable function such that $E(q(\xi_0))^2 < \infty$ and

$$D(\sum_{t \in I} q(\xi_t)) \geq C |I| , \quad C > 0, \quad I \in W. \qquad (9.4.3)$$

Then for this Gibbs random field and the sequence of ν -dimensional cubes $I_n = [-n, n]^\nu$, $n \in \mathbb{N}$, the c.l.th. is fulfilled.

Remark 9.4.1. Taking the function $q(x)$ as a potential ($q(x) = 0$, $x \in X_I$, $|I| > 1$) and using Theorem 9.3.1 we can reformulate the condition (9.4.3) of Theorem 9.4.1 in the following way: Let $q(x)$, $x \in X$, be a non-degenerated potential such that $E(q(\xi_0))^2 < \infty$.

Remark 9.4.2. Using Theorem 7.2.2. one can obviously obtain the c.l.th. conditions for Gibbs random field and the sequences tending to infinity in a sense of Van Hove.

Theorem 9.4.2. Let ξ_t, $t \in \mathbb{Z}^\nu$, be a centered Gibbs random field with values in some compact metric space X and let the corresponding potential Φ be such that

$$\sum_{J:0 \in J \in W} |J| (D(J))^\gamma \sup_{x \in X_J} |\Phi(x)| < \infty, \quad \gamma > 2 ,$$

$$\tfrac{1}{2} e^{4\|\Phi\|} (e^{4\|\Phi\|} - 1) < 1 .$$

Let for some $\delta > 0$ and some measurable function

$$E(q(\xi_0))^{2+\delta} < \infty, \quad D(\sum_{t \in I} q(\xi_t)) \geq C |I|, \quad C > 0, \quad I \in W.$$

Then for this Gibbs random field the law of the iterated logarithm is valid.

<u>Proof of Theorem 9.4.1.</u> According to Theorem 9.1.1 the φ -mixing coefficient for Gibbs random fields has the following form:

$$\varphi(\mathcal{F}_I, \mathcal{F}_V) = C_\Phi \sum_{k \geq 2} \left(\sum_{s_1 \in I} \sum_{s_2 \in \mathbb{Z}^\nu \setminus I} \cdots \sum_{s_k \in V} \tau(s_1, s_2) \times \right.$$

$$\left. \tau(s_2, s_3) \cdots \tau(s_{k-1}, s_k) \right),$$

where

$$\tau(a, b) = \exp\left\{ 4 \sum_{J: a, b \in J \in W} \sup_{x \in X_J} |\Phi(x)| \right\} - 1 , \quad a, b \in \mathbb{Z}^\nu.$$

Let

$$d(I, V) = \inf_{s_1 \in V_1, s_2 \in V_2} |s_1 - s_2|, \quad I, V \in W.$$

Then if $d(I, V) \geq d$ we can write

$$\varphi(\mathcal{F}_I, \mathcal{F}_V) \leq \sum_{s_1 \in I} \sum_{s_2 \in \mathbb{Z}^\nu \setminus \{s_1\}} \sum_{s_3 \in \mathbb{Z}^\nu \setminus \{s_2\}} \cdots \sum_{s_k \in V} \tau_{s_1 s_2} \tau_{s_2 s_3} \cdots \tau_{s_{k-1} s_k},$$

$$|s_1 + s_2 + \cdots + s_k| \geq d .$$

Since $|s_1 + s_2 + \cdots + s_k| \geq d$ there exists a number K_0 , $2 \leq K_0 \leq k$, such that $|s_{k_0} - s_{k_0-1}| \geq \frac{d}{k}$ and we find

$$\varphi(\mathcal{F}_I, \mathcal{F}_V) \leq |I| \sum_{k=2}^{\infty} \sum_{i=2}^{k} \left(\sum_{s_2 \in \mathbb{Z}^\nu \setminus \{s_1\}} \tau_{s_1 s_2} \cdots \sum_{\substack{s_i \in \mathbb{Z}^\nu \setminus \{s_{i-1}\} \\ |s_i - s_{i-1}| \geq \frac{d}{k}}} \tau_{s_{i-1} s_i} \cdots \right.$$

$$\left. \cdots \sum_{s_k \in \mathbb{Z}^\nu \setminus \{s_{k-1}\}} \tau_{s_{k-1} s_k} \right) .$$

Taking into account the condition (9.4.1)   we have

$$\sum_{u \in \mathbb{Z}^\nu \setminus \{s\}} \tau_{su} \leq \alpha < 1.$$

Thus

$$\varphi(\mathcal{B}_I, \mathcal{B}_v) \leq |I| \sum_{k=2}^{\infty} k \Big(\sum_{\substack{s \in \mathbb{Z}^{\nu} \setminus \{u\} \\ |s-u| \geq \frac{d}{k}}} \tau_{us} \Big) \Big(\sum_{s \in \mathbb{Z}^{\nu} \setminus \{u\}} \tau_{us} \Big)^{k-2} =$$

$$= |I| \Big(\sum_{k=2}^{d_0} k \Big(\sum_{\substack{s \in \mathbb{Z}^{\nu} \setminus \{u\} \\ |s-u| \geq \frac{d}{k}}} \tau_{us} \Big) \alpha^{k-2} + \sum_{k=d_0+1}^{\infty} k \alpha^{k-1} \Big)$$

By the condition (9.4.2) of Theorem 9.4.1 we find

$$\sum_{\substack{s \in \mathbb{Z}^{\nu} \setminus \{u\} \\ |s-u| \geq a}} \tau_{su} \leq \frac{1}{2} e^{4\|\Phi\|} \cdot 4 \sum_{\substack{s \in \mathbb{Z}^{\nu} \setminus \{u\} \\ |s-u| \geq a}} \sup_{x \in X_s, z \in X_u} |U_{\{s,u\}}^{\overline{y}}(x,z)| \leq$$

$$\leq 2 e^{4\|\Phi\|} \sum_{\substack{0 \in J \in W \\ (\mathrm{Diam}J) \geq a}} |J| \sup_{x \in X_J} |\Phi(x)| \leq \frac{2 e^{4\|\Phi\|} \|\Phi\|_{\gamma}}{a^{\gamma}}$$

$$\|\Phi\|_{\gamma} = \sum_{0 \in J \in W} |J| (\mathrm{Diam}\, J)^{\gamma} \sup_{x \in X_J} |\Phi(x)| < \infty$$

Hence

$$\varphi(\mathcal{B}_I, \mathcal{B}_v) \leq |I| \Big(\sum_{\substack{s \in \mathbb{Z}^{\nu} \setminus \{u\} \\ |s-u| \geq \frac{d}{d_0}}} \sum_{k=2}^{\infty} k \alpha^{k-2} + \sum_{k=d_0+1}^{\infty} k \alpha^{k-1} \Big) \leq$$

$$\leq |I| \Big(\|\Phi\|_{\gamma} \cdot 2 e^{4\|\Phi\|} \sum_{k=2}^{\infty} k \alpha^{k-2} \Big(\frac{d_0}{d} \Big)^{\gamma} + \sum_{k=d_0+1}^{\infty} k \alpha^{k-1} \Big)$$

Since

$$\sum_{k=2}^{\infty} k \alpha^{k-2} = \frac{2-\alpha}{(1-\alpha)^2} \ , \quad \sum_{k=d_0+1}^{\infty} k \alpha^{k-1} = \frac{\alpha^{d_0}(1+d_0-d_0\alpha)}{(1-\alpha)^2}$$

we have

$$\varphi(\mathcal{B}_I, \mathcal{B}_v) \leq |I| \Big(\frac{(2-\alpha)\|\Phi\|_{\gamma} 2 e^{4\|\Phi\|}}{(1-\alpha)^2} \Big(\frac{d_0}{d} \Big)^{\gamma} + \frac{\alpha^{d_0}(1+d_0-d_0\alpha)}{(1-\alpha)^2} \Big)$$

Putting $d_0 = \varepsilon \ln d$, $\varepsilon \ln \frac{1}{\alpha} < \gamma$, $\varepsilon > 0$ for $d \geq 2$ we obtain

$$\varphi(\mathcal{B}_I, \mathcal{B}_v) \leq \frac{|I|}{(1-\alpha)^2} \Big(\varepsilon^{\gamma} \cdot 2 e^{\|\Phi\|} (2-d) \|\Phi\|_{\gamma} \Big(\frac{\ln d}{d} \Big)^{\gamma} +$$

$$+ d^{\varepsilon \ln \alpha} (1 + \varepsilon(1-\alpha) \ln d) \leq |I| C_{\Phi} \Big(\frac{\ln d}{d} \Big)^{\gamma} , \quad \gamma > 0$$

and it remains to refer to Theorem 7.4.3.
Theorem 9.4.2 is proved similarly.

9.5 Cluster Properties and Mixing Conditions for Gibbs Random Fields with Vacuum Potential

In this section we consider the Gibbs random fields defined by means of the vacuum potential Φ . For simplicity we suppose that $X = \{\theta, a\}$ (the results in general case will be presented in Additions of this chapter). Thus the potential Φ is defined on the finite sets of the lattice $\mathbb{Z}^\nu$. Let

$$\| \Phi \| = \sum_{J : 0 \in J \in W} |\Phi(J)| < \infty . \qquad (9.5.1)$$

In statistical physics the next two parameters are usually separated: z ("activity") and $\beta > 0$ (inverse temperature). The Gibbs distribution in the finite volume Λ with the parameters z, β has the following form

$$(q_V)_I (x) = \frac{z^{|x|} \sum_{J \in \Lambda \setminus I} z^{|J|} \exp\{-\beta\, \mathcal{U}(J \cup x)\}}{\mathbb{Z}_\Lambda(\Phi, z, \beta)} , \qquad x \subset I \qquad (9.5.2)$$

$$\mathbb{Z}_\Lambda(\Phi, z, \beta) = \sum_{J \subset \Lambda} z^{|J|} \exp\{-\beta\, \mathcal{U}(J)\}.$$

In (9.5.2) we take empty boundary conditions, since in the sequel we will consider only the potentials for which the corresponding Gibbs random field is unique. It is also convenient for us to assume $\Phi(J) = 0$ if $|J| = 1$ setting that the activity includes $e^{\Phi(J)}$ for one-point subsets J . Here we will describe the Gibbs random fields with the help of so-called correlation functions. The latter are defined in the following way

$$\rho_\Lambda(x) = \mathbb{Z}_\Lambda^{-1}(\Phi, z, \beta) z^{|x|} \sum_{J \subset \Lambda \setminus x} z^{|J|} \exp\{-\beta\, \mathcal{U}(J \cup x)\}, \qquad (9.5.3)$$
$$x \subset \Lambda .$$

It is not difficult to verify that

$$(q_\Lambda)_I (x) = \sum_{\tilde{J} \subset I \setminus x} (-1)^{|\tilde{J}|} \rho(x \cup \tilde{J}), \qquad x \subset I . \qquad (9.5.4)$$

Let's denote

$$W(x) = \sum_{J : u \in J \subset x} \Phi(J), \qquad u \in x , \qquad (9.5.5)$$

$$W(x;y) = \sum_{J:\, u\in J\subset x} \Phi(J\cup y), \qquad\qquad (9.5.6)$$

$$K(x;y) = \sum_{n\geq 1}\ \sum_{\{J_1,\dots,J_n\}_y}\ \prod_{i=1}^{n}(e^{-W(x,J_i)}-1), \quad y\neq\varnothing. \qquad (9.5.7)$$

In (9.5.7) the summation is taken over all collections $\{J_1,J_2,\dots,J_n\}_y$ of the non-empty subsets of y such that $\bigcup_i J_i = y$. Note that

$$\sum_{y\subset \mathbb{Z}^{\nu}\setminus x} |K(x,y)| \leq 2\exp(e^{\|\Phi\|}-1).$$

Correlation functions (9.5.3) satisfy the so-called correlation equations (see /104/):

$$\rho_\Lambda(x) = \chi_\Lambda(x)\,\frac{z\,e^{-\beta W(x)}}{1+z\,e^{-\beta W(x)}}\Big[\rho_\Lambda(x') + \sum_{J:\, J\cap x=\varnothing} K(x,J)\big(\rho_\Lambda(x'\cup J) -$$
$$-\rho_\Lambda(x\cup J)\big)\Big], \qquad\qquad (9.5.8)$$

where

$$x' = x\setminus\{u\}, \quad u\in x, \quad \chi(x) = \begin{cases} 1, & x\subset\Lambda, \\ 0, & x\not\subset\Lambda. \end{cases}$$

Denote by E the Banach space of the bounded complex functions defined on $W = \{J\subset\mathbb{Z}^{\nu} : |J| < \infty\}$ with the uniform norm. Define the operator K acting in E by the formula

$$(K\varphi)(x) = \frac{z\,e^{-\beta W(x)}}{1+z\,e^{-\beta W(x)}}\Big[\varphi(x') + \sum_{J:\, J\cap x=\varnothing} K(x,J)(\varphi(x'\cup J) - \varphi(x\cup J))\Big].$$
$$(9.5.9)$$

Here if $|x| = 1$ then $\varphi(x') = 0$. Denoting

$$C = \sum_{J:\, 0\in J\in W} \Phi(J), \qquad D = \sum_{J:\, 0\in J\in W} |\Phi(J)|$$

one can obtain the following estimate

$$\|K\| < \left|\frac{z\,e^{\beta(D-C)}}{1+z\,e^{\beta(D-C)}}\right|(2\exp(e^{\beta D}-1)-1). \qquad\qquad (9.5.10)$$

<u>Theorem 9.5.1 (/40/ ,/114/).</u> Let

$$\left|\frac{z\,e^{\beta(D-C)}}{1+z\,e^{\beta(D-C)}}\right|(2\exp(e^{\beta D}-1)-1) < 1. \qquad\qquad (9.5.11)$$

Then in the complex domain $\mathcal{Z}$ defined by (9.5.11) the vector ρ_Λ is the unique solution in E of the following equation

$$\rho_\Lambda = \chi_\Lambda \frac{z}{1+z}\,\alpha + \chi_\Lambda\,K\,\rho_\Lambda\ ,$$

where $\alpha(x) = 1$ if $|x| = 1$ and $\alpha(x) = 0$ otherwise. For any $\varepsilon > 0$ there exists a finite set $\Delta \subset \mathbb{Z}^\nu$ such that

$$\left|\,\rho_\Lambda(x) - \rho(x)\,\right| < \varepsilon$$

if $x + \Delta \subset \Lambda$ and ρ is the unique solution of the equation

$$\rho = \frac{z}{1+z}\,\alpha + K\rho\ .$$

Thus for sufficiently small $\|\Phi\|$ or sufficiently small z there exist the unique limiting correlation functions ρ and hence according to (9.5.4) the limiting finite-dimensional densities

$$q_I(x) = \lim_{\Lambda \to \infty}(q_\Lambda)_I(x),\quad I \in W,\quad x \subset I$$

are consistent in a sense of Kolmogorov and define the unique Gibbs random field with given potential.

Further it will be technically convenient to introduce so-called "non-physical" configurations in Λ as the finite sequences
$x = (x_1, x_2, ..., x_n)$, $x_i \in \Lambda$, $i = \overline{1,n}$. Note that usual (physical) configurations correspond to the case when all x_i, $i = \overline{1,n}$ are distinct. Denote

$$n(x) = N_1!\, ...\, N_K!\ ,$$

when x_i, $i = \overline{1,n}$, occupy only k different positions occuring respectively $N_1, ..., N_K$ times.
Now let's introduce some necessary definitions.
By generalized graph $G = (J_1, ..., J_K)$, $k \in \mathbb{N}$, with vertexes in some configuration T we will mean an arbitrary finite sequence of configuration $J_i \subseteq T$, $i = \overline{1,k}$ any of which contains at least two points. For the ordinary graphs we have

$$|J_i| = 2,\quad i = \overline{1,k}\ .$$

The generalized graph is called connected with respect to two points $t_1, t_2 \in T$ if there exists a sequence $\{J_{i_1}, ..., J_{i_S}\}$, $J_{i_m} \in G$, $m = \overline{1,S}$ such that $J_{i_m} \cap J_{i_{m+1}} \neq \varnothing$, $m = \overline{1,S-1}$ and $t_1 \in J_{i_1}$, $t_2 \in J_{i_S}$.

We say that the generalized graph is connected with respect to some configuration $V \subseteq T$ if it is connected with respect to any two points $t_1, t_2 \in V$. Let $V_1, V_2 \subset T$ be two configurations. It is said that generalized graph is connected with respect to V_1 and all points V_2 if the graph $G' = (J_1, ..., J_K, V_1)$ is connected with respect to configuration V_2 . Further if

$$x = (x_1, ..., x_N), \quad y = (y_1, ..., y_N) \qquad \text{by} \quad (xy) \qquad \text{we denote the}$$

configuration $(x_1, ..., x_N, y_1, ..., y_N)$. We put below $\Phi(u, u) = +\infty$, $u \in \mathbb{Z}^\nu$ and let $\Phi(u, u, v_1, ..., v_p)$, $p \geq 1$, $v_i \in \mathbb{Z}^\nu$, be finite and arbitrary. Now let's note that if z belongs to the domain (9.5.11) then the correlation function has the following representation (see /38/)

$$\rho_\Lambda(x) = \begin{cases} \sum\limits_{n \geq 0} z^n (n!)^{-1} \sum\limits_{y \in \Lambda^{(n)}} \varphi(x; y), & n(x) = 1, \\ 0, & n(x) > 1. \end{cases}$$

where the coefficients $\varphi(x; y)$ are called the Ursell functions and may be defined by the formula

$$\varphi(x; y) = \sum\limits_{G \in \Gamma_c(x; y)} \prod\limits_{J \in G} \left(e^{-\Phi(J)} - 1 \right). \qquad (9.5.12)$$

Here $\Gamma_c(x; y)$ is the set of all generalized graphs connected with respect to x and all points y . The Ursell functions $\varphi(x; y)$ satisfy the next induction relations

$$\varphi(x; y) = e^{-W(x)} \sum\limits_{\widetilde{y} \subset y} K(x; \widetilde{y}) \, \varphi(x'\widetilde{y}; y \setminus \widetilde{y}). \qquad (9.5.13)$$

The decay of correlations may be represented in terms of so-called truncated correlation functions ρ_Λ^T :

$$\rho_\Lambda^T(x) = \rho_\Lambda(x) - \sum\limits_{\{J_1, J_2, ..., J_i\}} \prod\limits_{s=1} \rho_\Lambda^T(J_s), \qquad (9.5.14)$$

$$\rho_\Lambda^T(x_1) = \rho_\Lambda(x_1),$$

where the sum runs over all partitions $\{J_1, J_2, ..., J_i\}$ of

$$x = (x_1, ..., x_N), \quad N > 1.$$

The truncated function has in the domain (9.5.4) the next representation

$$\rho_\Lambda^T(x) = \sum_{n \geqslant 0} z^n (n!)^{-1} \sum_{y \in \Lambda^n} \varphi(x,y).$$ (9.5.15)

For any translation invariant metric d on $\mathbb{Z}^\nu$ let's introduce a new potential Φ_d :

$$\Phi_d(J) = e^{L_d(J)} |\Phi(J)|,$$

where $L_d(J)$ is the minimal length (with respect to the metric d) of all the trees the vertexes set of which contains all points of J and perhaps some other ones. We will consider potentials Φ satisfying the condition

$$D_d = \sum_{0 \in J \subset \mathbb{Z}^\nu} e^{L_d(J)} |\Phi(J)| < \infty.$$ (9.5.16)

Note that in the case $d(t,t') = a|t-t'|$, $t,t' \in \mathbb{Z}^\nu$ or $d(t,t') = s \log(1+a|t-t'|)$, $a > 0$, $t,t' \in \mathbb{Z}^\nu$ (9.5.16) means the exponential or power decay of potential, respectively.

<u>Theorem 9.5.2 (/38/).</u> Let

$$|z| < C_d^{-1}, \qquad C_d = 2 \exp D_d \exp(\exp D_d - 1).$$

Then

$$|\rho_\Lambda^T(x)| \leqslant C_d^{-1} \left(\frac{z C_d}{1 - z C_d}\right)^{|x|} n(x) e^{-L_d(x)}.$$ (9.5.17)

(9.5.17) is a strong cluster property if $d(t,t')$ increases faster than $(\nu + \varepsilon) \log(1+a|t-t'|)$ with $\varepsilon > 0$ and $a > 0$.

<u>Proof.</u> From the definitons of the corresponding expresson and (9.5.16) it follows that

$$|W(x;y)| \leqslant \sum_{J:(u,J) \subset B \subset (x,y)} e^{-L_d(B)} \Phi_d(B) \leqslant$$

$$\leqslant e^{-L_d(u,Y)} W_d(x;y).$$

Using the inequality

$$e^{ax} - 1 \leqslant x(e^a - 1) \quad \text{where} \quad x \in [0,1], \ a \geqslant 0 \qquad \text{we obtain}$$

$$e^{|W(x;y)|} - 1 \leqslant e^{-L_d(u,y)}\left(e^{W_d(x;y)} - 1\right). \tag{9.5.18}$$

Hence

$$|K(x;y)| \leqslant \sum_{k\geqslant 1} \sum_{\{J_1,\ldots,J_k\}_y} \prod_{i=1}^{k} e^{-L_d(u;J_i)}\left(e^{W_d(x;J_i)} - 1\right) \leqslant$$

$$\leqslant e^{-L_d(u,y)} K_d(x;y).$$

Since

$$K(x;y) = -K(x;(u,y))$$

then

$$|K(x;y)| \leqslant e^{-L_d(u,y)} K_d(x;(u,y))$$

where

$$K_d(x;y) = \begin{cases} \sum_{k\geqslant 1} \sum_{\{J_1,\ldots,J_k\}} \prod_{i=1}^{k} \left(e^{W_d(x;J_i)} - 1\right), & y \subset \mathbb{Z}^\nu \setminus x \\ 0, & y \subset \mathbb{Z}^\nu \setminus x' \end{cases}$$

and

$$\sum_{y \subset \mathbb{Z}^\nu \setminus x'} K_d(x;y) \leqslant 2\exp(\exp D_d - 1).$$

Lemma 9.5.1. For the function $\varphi_d(x;y)$ defined by the inductive relations

$$\varphi_d(x;y) = e^D \sum_{\tilde{y} \subset x} K_d(x;\tilde{y})\,\varphi_d(x'\tilde{y}; y \setminus \tilde{y}),$$

$$\varphi_d(\{u\}) = 1, \quad u \in \mathbb{Z}^\nu,$$

the following estimate is valid:

$$|\varphi(x;y)| \leqslant e^{-L_d(x;y)} \varphi_d(x;y), \tag{9.5.19}$$

where $L_d(x;y)$ is the minimum with respect to the metric d of the trees length constructed on the points of x, y and perhaps arbitrary other ones which are connected with respect to the cluster x and points of y .

<u>Proof.</u>   We will use the induction by the sum $|x|+|y|=n$. Let for all x, y such that $|x|+|y| < n$ (9.5.19) be fulfilled. Taking into account (9.5.13) and (9.5.18) we find

$$|\varphi(x;y)| \le e^{D} \sum_{\tilde{y} \subset y} e^{-L_d(x,y)} K_d(\dot{x};\tilde{y}) \varphi(x\tilde{y}, y \smallsetminus \tilde{y}) \le$$

$$\le e^{D} \sum_{\tilde{y} \subset y} e^{-L_d(x,y)} K_d(x;y) e^{-L_d(x'\tilde{y}; y \smallsetminus \tilde{y})} \varphi_d(x'\tilde{y}; y \smallsetminus \tilde{y}).$$

But

$$L_d(u,y) + L_d(x'\tilde{y}; y \smallsetminus \tilde{y}) \ge L_d(x;y)$$

and hence the lemma is proven.

<u>Lemma 9.5.2.</u> For any $n \in \mathbb{N}$ and configuration x the following estimate is valid:

$$\sum_{y \in \mathbb{Z}^{\nu n}} \varphi_d(x;y) \le C_d^{|x|+n-1} \cdot n! \qquad (9.5.20)$$

<u>Proof.</u>   Again we use the induction by sum $|x| + |y|$. We have

$$\sum_{y \in \mathbb{Z}^{\nu n}} \varphi_d(x;y) = e^{D} \sum_{y \in \mathbb{Z}^{\nu n}} \sum_{\tilde{y} \subset y} K_d(x;\tilde{y}) \varphi_d(x'\tilde{y}; y \smallsetminus \tilde{y}) =$$

$$= e^{D} \sum_{\tilde{y} \in \mathbb{Z}^{\nu n}} K_d(x;\tilde{y}) \sum_{y:\tilde{y} \subset y \in \mathbb{Z}^{\nu n}} \varphi_d(x'\tilde{y}; y \smallsetminus \tilde{y}) =$$

$$= e^{D} \sum_{\tilde{y} \in \mathbb{Z}^{\nu n}} K_d(x;\tilde{y}) \sum_{y \in \mathbb{Z}^{\nu(n-|\tilde{y}|)}} \varphi_d(x'\tilde{y}; y) \le$$

$$\le e^{D} \sum_{\tilde{y} \in \mathbb{Z}^{\nu n}} K_d(x;\tilde{y}) C_d^{|x'|+|\tilde{y}|+n-|\tilde{y}|-1} (n-|\tilde{y}|)! \le$$

$$\leq C_d^{|x|+n-1} \cdot n!$$

The proof is finished.

Further according to (9.5.15)   we have

$$\left| \rho_\Lambda^T(x) \right| \leq |z|^{|x|} \sum_{n \geq 0} \frac{|z|^n}{n!} \sum_{y \in \Lambda^n} |\varphi(x,y)|.$$

Using the relation (see /114/)

$$\varphi(x,y) = \varphi(u; x'y), \qquad u \in \mathbb{Z}^\nu, \qquad x' = x \smallsetminus u$$

we obtain

$$\left| \rho_\Lambda^T(x) \right| \leq |z|^{|x|} \sum_{n \geq 0} \frac{|z|^n}{n!} \sum_{y \in \Lambda^n} |\varphi(u; x'y)| \leq$$

$$\leq |z|^{|x|} \sum_{n \geq 0} \frac{|z|^n}{n!} \sum_{y \in \Lambda^n} e^{-L_d(u; x'y)} \varphi_d(u; x'y) \leq$$

$$\leq |z|^{|x|} e^{-L_d(x)} \sum_{n \geq 0} \frac{|z|^n}{n!} \sum_{y \in \Lambda^{n+|x|-1}} \varphi_d(u; y) \leq$$

$$\leq |z|^{|x|} e^{-L_d(x)} C_d^{|x|-1} \sum_{n \geq 0} \frac{(|x|+n-1)! \, C_d^n}{n!} \cdot |z|^n$$

from which follows the theorem.

Note, however, the estimate obtained in Theorem 9.5.2 is not convenient for the application. The estimate expressed in terms of paths is much more convenient. The following lemma presents such an estimate.

<u>Lemma 9.5.3 (/38/)</u>. Let $\mathcal{L}(x)$ (correspondingly $\mathcal{E}(x)$) be the set of all trees (correspondingly paths) constructed on the points of x and let $L_d(T)$ (correspondingly $L_d(C)$) be the length of tree T (correspondingly of path C) with respect to metric d .   Then

$$\sum_{T \in \mathcal{L}(x)} e^{-L_\delta(T)} \leq \frac{2|x|^{|x|-2}}{|x|!} \sum_{C \in \mathcal{E}(x)} e^{-\frac{1}{2} L_\delta(C)}.$$

<u>Proof.</u> To each tree $T \in \mathcal{L}(x)$ corresponds the set of paths $\mathcal{E}_T$ by the following rule: Starting from arbitrary point of the tree and "turning around" T keeping each point $x_i \in x$ once and only once as soon as we meet it. Duneau and Souillard demonstrate this rule by means of the next picture:

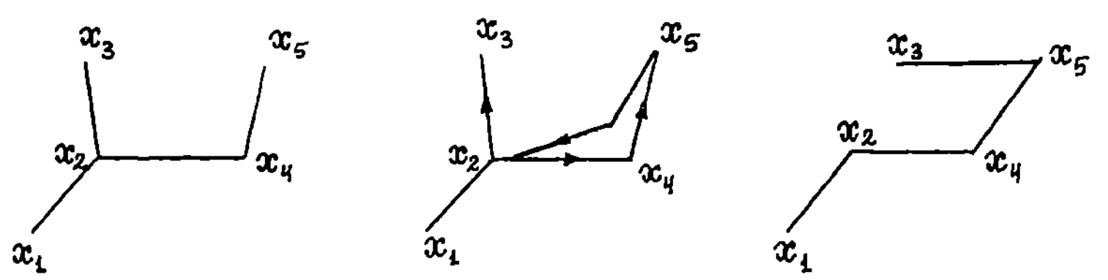

Tree   Path

Then

$$2 L_d(T) > L_d(C)$$

for any $C \in \mathcal{E}_T$.

Let $v(T)$ be the number of distinct paths associated with the tree T. Then we have

$$\sum_{T \in \mathcal{L}(x)} e^{-L_d(T)} = \sum_{C \in \mathcal{E}(x)} \sum_{T : C \in \mathcal{E}_T} e^{-L_d(T)} v(T)^{-1} \leq$$

$$\leq \sum_{C \in \mathcal{E}(x)} e^{-\frac{1}{2} L_d(C)} \sum_{T : C \in \mathcal{E}_T} v(T)^{-1}.$$

Of course if $\delta = 0$ then the latter represents an equation. Thus by the symmetry of last factor we obtain

$$\sum_{T \in \mathcal{L}(x)} 1 = |x|^{|x|-2} = \sum_{C \in \mathcal{E}(x)} \sum_{T : C \in \mathcal{E}_T} v(T)^{-1}$$

and hence

$$\sum_{T : C \in \mathcal{E}_T} v(T)^{-1} = \frac{2|x|^{|x|-2}}{|x|!}.$$

__Theorem 9.5.3__ Let a potential Φ be such that for some translation-invariant metric d

$$\left| \frac{z\, C_d\, e^{\|\Phi\|+1}}{1 - z\, C_d} \right| \sum_{t:\, t\in\mathbb{Z}^{\nu}\setminus\{0\}} e^{-\frac{1}{2} d(0,t)} \leqslant \alpha < 1$$

and $\quad |z| < C_d^{-1}$

Then

$$\varphi(\mathcal{F}_I, \mathcal{F}_V) \leqslant C(\Phi, \mu, z, \beta)\, |I| \sum_{t\in\mathbb{Z}^{\nu}:\, |t| \geqslant \iota(I,V)} e^{-\frac{1}{2} d(0,t)},$$

$$I \wedge V = \varnothing, \quad I, V \in W, \qquad \iota(I,V) = \min_{t\in I,\, s\in V} \{|t-s|\}$$

__Proof__ Using (9.5.4.) we have

$$A_{I,V}(x,y) = (q_\Lambda)_{I\cup V}(x\cup y) - (q_\Lambda)_I(x)(q_\Lambda)_V(y) =$$

$$= \sum_{\tilde{x}:\, x\subset\tilde{x}\subset I}\ \sum_{\tilde{y}:\, y\subset\tilde{y}\subset V} (-1)^{|\tilde{x}\setminus x| + |\tilde{y}\setminus y|} \left(\rho_\Lambda(\tilde{x}\cup\tilde{y}) - \rho_\Lambda(\tilde{x})\rho_\Lambda(\tilde{y}) \right) \tag{9.5.20'}$$

According to (9.5.14) we can write

$$\rho_\Lambda(x) = \sum_{J:\, a\in J\subset x} \rho_\Lambda^T(J)\, \rho_\Lambda(x\setminus J)$$

Using this relation we obtain

$$\rho_\Lambda(x\cup y) = \sum_{J:\, a\in J\subset x}\ \sum_{J':\, J'\subset y} \rho_\Lambda^T(J\cup J')\, \rho_\Lambda(x\setminus J \cup y\setminus J')$$

$$\rho_\Lambda(x)\, \rho_\Lambda(y) = \sum_{J:\, a\in J\subset x} \rho_\Lambda^T(J)\, \rho_\Lambda(x\setminus J)\, \rho_\Lambda(y),$$

$$\rho_\Lambda(x \cup y) - \rho_\Lambda(x)\,\rho_\Lambda(y) = \sum_{J:\,a\in J\subset x} \rho_\Lambda^T(J)\,(\rho_\Lambda(x\setminus J\cup y) - $$

$$- \rho_\Lambda(x\setminus J)\,\rho_\Lambda(y)) + \sum_{J_i':\,a\in J_i'\subset x}\ \sum_{J_i'':\,\emptyset\neq J_i''\subset y} \rho_\Lambda^T(J_i'\cup J_i'')\,\rho_\Lambda(x\setminus J_i'\cup y\setminus J_i'') \qquad (9.5.20'')$$

Substituting $(9.5.20'')$ into $(9.5.20')$ we have the following recursive relation

$$A_{I,V}(x,y) = \sum_{\widetilde{x}:\,x\subset\widetilde{x}\subset I}(-1)^{|\widetilde{x}\setminus x|} \sum_{\widetilde{y}:\,y\subset\widetilde{y}\subset V}(-1)^{|\widetilde{y}\setminus y|} \sum_{J_i':\,a\in J_i'\subset\widetilde{x}}\ \sum_{J_i'':\,\emptyset\neq J_i''\subset\widetilde{y}}$$

$$\rho_\Lambda^T(J_i'\cup J_i'')\,\rho_\Lambda(\widetilde{x}\setminus J_i\cup\widetilde{y}\setminus J_i'') + \sum_{J:\,a\in J\subset J}(-1)^{|J|}\rho_\Lambda^T(J)\,A_{I\setminus J}(x\setminus J,y) =$$

$$= \sum_{J_i':\,a\in J_i'\subset I}\ \sum_{J_i'':\,\emptyset\neq J_i''\subset V}\rho_\Lambda^T(J_i'\cup J_i'')\sum_{\substack{\widetilde{x}:\,x\subset\widetilde{x}\subset I\\ J_i'\subset\widetilde{x}}}(-1)^{|\widetilde{x}\setminus x|}\sum_{\substack{\widetilde{y}:\,y\subset\widetilde{y}\subset V\\ J_i''\subset\widetilde{y}}}(-1)^{|\widetilde{y}\setminus y|}$$

$$\rho_\Lambda(\widetilde{x}\setminus J_i\cup\widetilde{y}\setminus J_i'') + \sum_{J:\,a\in J\subset I}(-1)^{|J|}\rho_\Lambda^T(J)\,A_{I\setminus J}(x\setminus J,y)$$

Hence

$$A_{I,V}(x,y) = \sum_{J_i':\,a\in J_i'\subset I}\ \sum_{J_i'':\,\emptyset\neq J_i''\subset V}\rho_\Lambda^T(J_i'\cup J_i'')(-1)^{|J_i'\setminus x|+|J_i''\setminus y|}\times \qquad (9.5.21)$$

$$\times(q_\Lambda)_{I\cup V\setminus(J_i'\cup J_i'')}(x\setminus J_i'\cup y\setminus J_i'') + \sum_{J:\,a\in J\subset I}(-1)^{|J|}\rho_\Lambda^T(J)\,A_{I\setminus J}(x\setminus J,y)$$

Let us now estimate the following ratio

$$\frac{(q_\Lambda)_{I\cup V\setminus(J_i'\cup J_i'')}(x\setminus J_i'\cup y\setminus J_i'')}{(q_\Lambda)_I(x)} =$$

$$= \frac{\sum\limits_{J \subset (\Lambda \smallsetminus (I \cup V)) \cup J_i' \cup J_i''} \exp\left\{-U(x \smallsetminus J_i' \cup y \smallsetminus J_i'' \cup J)\right\}}{\sum\limits_{z \subset \Lambda \smallsetminus I} \exp\left\{-U(x \cup J)\right\}} =$$

$$= \frac{\sum\limits_{z' \subset J_i'} \sum\limits_{z'' \subset J_i''} \sum\limits_{J \subset \Lambda \smallsetminus (I \cup V)} \exp\left\{-U(x \smallsetminus J_i' \cup y \smallsetminus J_i'' \cup z' \cup z'' \cup J)\right\}}{\sum\limits_{z \subset V} \sum\limits_{J \subset \Lambda \smallsetminus (I \cup V)} \exp\left\{-U(x \cup z \cup J)\right\}} \leq$$

$$\leq \frac{\sum\limits_{z' \subset J_i'} \sum\limits_{z \subset V} \sum\limits_{J \subset \Lambda \smallsetminus (I \cup V)} \exp\left\{-U(x \smallsetminus J_i' \cup z' \cup z \cup J)\right\}}{\sum\limits_{z \subset V} \sum\limits_{J \subset \Lambda \smallsetminus (I \cup V)} \exp\left\{-U(x \cup z \cup J)\right\}} =$$

$$= \left(\sum\limits_{z \subset V} \sum\limits_{J \subset \Lambda \smallsetminus (I \cup V)} \exp\left\{-U(x \cup z \cup J)\right\}\right)^{-1} \times$$

$$\times \sum\limits_{z' \subset J_i'} \sum\limits_{z \subset V} \sum\limits_{J \subset \Lambda \smallsetminus (I \cup V)} \exp\left\{-U(x \cup z \cup J)\right\} \exp\left\{-U(x \smallsetminus J_i' \cup z' \cup z \cup J)\right. +$$

$$\left. + U(x \cup z \cup J)\right\}.$$

Since

$$\left| U(x \smallsetminus J_i' \cup z' \cup z \cup J) - U(x \cup z \cup J) \right| \leq \|\Phi\| \, |J_i'|$$

then we have

$$(q_\Lambda)_{I \cup V \smallsetminus (J_i' \cup J_i'')} (x \smallsetminus J_i' \cup y \smallsetminus J_i'') \leq e^{\|\Phi\| \, |J_i'|} (q_\Lambda)_I (x).$$

Substituting this inequality into (9.5.21) we obtain

$$\left| A_{I,V}(x,y) \right| \leq \sum\limits_{J_i' : \varnothing \neq J_i' \subset I} \sum\limits_{J_i'' : \varnothing \neq J_i'' \subset V} \left| \rho_\Lambda^T (J_i' \cup J_i'') \right| e^{\|\Phi\| \, |J_i'|} (q_\Lambda)_I (x) +$$

$$+ \sum_{J:\, a\in J\subset I} (-1)^{|J|}\, \rho_\Lambda^T(J)\, A_{I\setminus J}(x\setminus J, y) \qquad\qquad (9.5.21')$$

Let us now use Theorem 9.5.3 and Lemma 9.5.3. We can write

$$\sum_{\substack{J:\, a\in J\subset I\cup V \\ a\in I,\, J\cap V\neq\varnothing}} e^{\|\Phi\|\,|J|}\, \rho_\Lambda^T(J) \leqslant C_d^{-1} \sum_{\substack{J:\, a\in J\subset I\cup V \\ a\in I,\, J\cap V\neq\varnothing}} \left|\frac{z\,C_d\,e^{\|\Phi\|}}{1-z\,C_d}\right|^{|J|} e^{-L_d(J)} \leqslant$$

$$\leqslant 2C_d^{-1} \sum_{\substack{J:\, a\in J\subset I\cup V \\ a\in I,\, J\cap V\neq\varnothing}} \left|\frac{z\,C_d\,e^{\|\Phi\|}}{1-z\,C_d}\right|^{|J|} \frac{|J|^{|J|-2}}{|J|!} \sum_{c\in\mathcal{E}(J)} e^{-\frac{1}{2}L_d(c)} \leqslant$$

$$\leqslant \sum_{\substack{J:\, a\in J\subset I\cup V \\ a\in I,\, J\cap V\neq\varnothing}} \beta^{|J|} \sum_{c\in\mathcal{E}(J)} e^{-\frac{1}{2}L_d(c)}$$

Here

$$\beta = \left|\frac{z\,C_d\,e^{\|\Phi\|+1}}{1-z\,C_d}\right|$$

Thus

$$\sum_{\substack{J:\, a\in J\subset I\cup V \\ a\in I,\, J\cap V\neq\varnothing}} e^{\|\Phi\|\,|J|}\, \rho_\Lambda^T(J) \leqslant$$

$$\leqslant \sum_{t\in V} \sum_{J\subset \mathbb{Z}^\nu\setminus\{a,t\}} \beta^{|J|} \sum_{c\in\mathcal{E}(J\cup\{a,t\})} e^{-\frac{1}{2}L_d(c)} \leqslant$$

$$\leqslant \sum_{t\in V} \sum_{n\geqslant 0} \frac{1}{n!} \sum_{J\in\mathbb{Z}^{\nu n}} \beta^n \sum_{(a,t)\in c\in\mathcal{E}(J)} e^{-\frac{1}{2}L_d(c)} \leqslant$$

$$\leqslant \sum_{t\in V} e^{-\frac{1}{2}d(a,t)} \sum_{n\geqslant 1} \alpha^n \leqslant$$

$$\leqslant \frac{\alpha}{1-\alpha} \sum_{\substack{t\in \mathbb{Z}^\nu \setminus \{0\} \\ |t|\geqslant \iota(I,V)}}$$

In exactly the same way one can show that

$$\sum_{J:a\in J\subset I} |\rho_\wedge^T(J)| < 1$$

Hence using the recursive relation (9.5.21')

$$|(q_\wedge)_{I\cup V}(x\cup y) - (q_\wedge)_I(x)(q_\wedge)_V(y)| \leqslant$$

$$\leqslant C^{(1)}(\Phi,\beta,z)\,|I| \sum_{t\in \mathbb{Z}^\nu:\,|t|\geqslant \iota(I,V)} e^{-\frac{1}{2}d(0,t)}$$

Thus

$$|(P_\wedge)_{I\cup V}(AB) - (P_\wedge)_I(A)(P_\wedge)_V(B)| \leqslant$$

$$\leqslant C^{(1)}(\Phi,\beta,z)\,\mu^2(x)\,|I| \sum_{t\in \mathbb{Z}^\nu:\,|t|\geqslant \iota(I,V)} e^{-\frac{1}{2}d(0,t)} (P_\wedge)_I(B),$$

$$A\in \mathcal{F}_I, \quad B\in \mathcal{F}_V.$$

Passing to the limit in the last relation as $\Lambda \to \infty$ we finish the proof of the theorem.

<u>Remark 9.5.1.</u> It is easy to see that the next relation also follows from the proof of Theorem 9.5.3:

$$\left| (P_\Lambda)_{I \cup V}(AB) - (P_\Lambda)_I(A)(P_\Lambda)_V(B) \right| \le$$

$$\le C^{(2)}(\Phi, d, z)\, \mu^2(X)\, |I||V|\, (P_\Lambda)_I(A)(P_\Lambda)_V(B) \times$$

$$\times \sup_{a \in I,\ b \in V} \left\{ e^{-\frac{1}{2} d(a,b)} \right\}.$$

Now using the obtained results we can present the limit theorems.

<u>Theorem 9.5.4.</u> Let ξ_t, $t \in \mathbb{Z}^\nu$, be a Gibbs random field with vacuum potential Φ and let

$$|z| < C_d^{-1}, \qquad \left| \frac{e^{\|\Phi\|+1} z\, C_d}{1 - z\, C_d} \right| \sum_{t \in \mathbb{Z}^\nu \setminus \{0\}} e^{-\frac{1}{2} d(0,t)} \le \alpha < 1,$$

$$\sum_{t \in \mathbb{Z}^\nu:\, |t| > d} e^{-\frac{1}{2} d(0,t)} \le \frac{1}{d^{2\nu + \varepsilon}}, \qquad \varepsilon > 0.$$

Let a measurable function $g(x)$, $x \in X$ be such that $Eg(\xi_0) = 0$, $Eg^2(\xi_0) < \infty$. Then the series

$$\sigma_g^2 = \sum_{t \in \mathbb{Z}^\nu} Eg(\xi_0) g(\xi_t)$$

converges and if $\sigma_g^2 \ne 0$ then the sequence $S_{I_n} = (DS_{I_n})^{-\frac{1}{2}} \sum_{t \in I_n} g(\xi_t)$, $I_n = [-n, n]^\nu$, $n \in \mathbb{N}$ is asymptotically normal.

<u>Theorem 9.5.5.</u> Under conditions of Theorem 9.5.4 if in addition for some $\delta > 0$ the moment $E|g(\xi_0)|^{2+\delta}$ exists then for the sequence S_{I_n} the law of the iterated logarithm is valid.

<u>9.6. Additions</u>

The correlation functions in general case we define by the following way

$$\rho_{\Lambda}(x) = \chi_{\Lambda}(x)\int_{X_{\Lambda\setminus I}} \exp\left\{-\mathcal{U}(x\,y)\right\}\mu_{\Lambda\setminus I}(dy), \quad x\in X_I^*, \ I\subset\Lambda, \quad X^*= X\setminus\{\theta\}$$

where

$$\chi_{\Lambda}(x) = \begin{cases} 1, & x\in X_I, \ I\subset\Lambda \\ 0 & \text{otherwise.} \end{cases}$$

As it was shown in /83/   these correlation functions satisfy the
equations

$$\rho_{\Lambda}(x) = \chi_{\Lambda}(x)\exp\left\{-W(x)\right\}\left(1+\int_{X^*}\exp\left\{-W(x';z)\right\}\mu(dz)\right)^{-1}\times$$

$$\times\left(\rho_{\Lambda}(x') - \int_{X^*}\exp\left\{-W(x';z)\right\}\left(G_{\Lambda}\rho_{\Lambda}\right)(x';z)\mu(dz) + \right.$$

$$\left. + \left(1+\int_{X^*}\exp\left\{-W(x';z)\right\}\mu^*(dz)\right)\left(G_{\Lambda}\rho_{\Lambda}\right)(x)\right).$$

Here

$$W(x) = \sum_{J\subset I}\Phi(x_J), \quad x\in X_I^*,$$

$$W(x;y) = \sum_{J\subset I}\Phi(x_J,y) \quad x\in X_I^*, \ y\in X_V^*,$$

$$I,V\in W, \quad I\cap V=\varnothing, \quad x' = x\setminus\{x_1\},$$

$$\left(G_{\Lambda}\rho_{\Lambda}\right)(x) = \sum_{J\in\Lambda\setminus I}\int_{X_J^*} K(x,y)\left(\rho_{\Lambda}(x'y) - \right.$$

$$- \int_{X_I^*} \rho_\Lambda (x' \, y \, z) \, \mu_t (dz)) \, \mu_J (dy)$$

$$K(x,y) = \sum_{n \geqslant 1} \; \sum_{\{J_1, \ldots, J_n\}} \; \prod_{j=1}^{n} (e^{-W(x, \, y_{J_j})} - 1),$$

where summation extended over all sets $\{J_1, \ldots, J_n\}$, $J_j \in W_v$, $J_j \neq \emptyset$, $i = 1, \ldots, n$ such that $\bigcup_j J_j = y$.

Using this definition for the potential Φ with norm

$$\| \Phi \| = \sum_{J : 0 \in J \in W} \; \sup_{x \in X_J} |\Phi(x)| < \infty$$

and sufficiently small activity

$$|z| < \left\{ 2 e^{\| \Phi' \|} \exp(e^{\| \Phi \|'} - 1) \right\}^{-1}$$

one can prove the existence of a unique Gibbs random field with such potential (see /83/).

The definition of the truncated correlation function is similar to

$$\rho_\Lambda^T (x) = \rho_\Lambda(x) - \sum_{\{J_1, \ldots, J_n\}} \; \prod_{s=1}^{n} \rho_\Lambda^T (x_{J_s}),$$

$$\rho_\Lambda^T (x_i) = \rho_\Lambda(x_i).$$

Note also that the analogy of Theorem 9.5.2 for spin systems was established in /2/ :

For any configuration $x \in X_I$ let us denote

$$\text{supp } x = \{ t \in I : x_t \neq \theta \}.$$

<u>Theorem 9.6.1.</u> Let a potential Φ be such that

$$\| \Phi \|_d = \sum_{J : 0 \in J \in W} e^{L_d(J)} \sup_{x \in X_J} |\Phi(x)| < \infty$$

and

$$\cdot |z| < C_d^{-1} = \left[2\, e^{\|\Phi'\|_d}\, \exp\left\{ e^{\|\Phi'\|} - 1 \right\} \right]^{-1}.$$

Then for any $\Lambda \subset \mathbb{Z}^\nu$

$$\left| \rho_\Lambda^T(x) \right| \leq |z|\, C_d^{-1} \left(\frac{C_d}{1 - C_d} \right)^{|\mathrm{supp}\, x|} e^{-L_d(\mathrm{supp}\, x)}.$$

Using the cluster estimates (9.5.17) Pogosian /108/ has obtained the asymptotical expansion of the logarithm of the partition function.

Denote $\mathscr{R}(\tau)$, $\tau > 0$, the class of parallelepipeds $\Lambda \subset \mathbb{Z}^\nu$ satisfying the condition:

$$\mathrm{diam}\, \Lambda \leq \tau\, |\Lambda|^{1/\nu}.$$

For any $\Lambda \in \mathscr{R}(\tau)$ denote by $\Lambda^{(i)}$, $i = 0,1,\ldots,\nu-1$, the family of all i -faces of Λ .

Potential Φ is called Euclidean invariant if Φ is invariant under the action of the group G of all automorphisms of $\mathbb{Z}^\nu$. A metric d is called Euclidean invariant if $d(u,t) = d(gu, gt)$ for all $g \in G$, $u, t \in \mathbb{Z}^\nu$.

Let us consider an Euclidean invariant potential Φ satisfying the conditions: there exists an Euclidean invariant metric d on $\mathbb{Z}^\nu$ such that

$$\sum_{t \in \mathbb{Z}^\nu} \exp\left\{ -\tfrac{1}{2} d(0,t) \right\} < \infty$$

and

$$\sum_{0 \in J \in W} |\Phi(J)|\, e^{L_d(J)} < \infty.$$

$$\text{(9.6.1)}$$

<u>Theorem 9.6.2.</u> Let the activity z be sufficiently small and let the potential Φ satisfy the conditions (9.6.1). Then for any $\Lambda \in \mathscr{R}(\tau)$ the following expansion holds:

$$\ln Z_\Lambda(\beta, z, \Phi) = a_0 |\Lambda| + a_1 \sum_{\lambda \in \Lambda^{(\nu-1)}} |\lambda| + \cdots + a_{\nu-1} \sum_{\lambda \in \Lambda^{(1)}} |\lambda| + a_\nu + R_\nu(\Lambda),$$

where a_i , $i = \overline{1,\nu}$ are some constants depending on the potential

Φ and the parameters β, z the quantity $R_\nu(\Lambda) = R_\nu(\Lambda, \Phi, \beta, z)$ satisfies the estimate:

$$|R_\nu(\Lambda)| \leqslant C |\Lambda|^{-\frac{1}{\nu}}, \quad 0 < C < \infty,$$

with the constant $C = C(\Phi, \beta, z, \nu)$. Moreover, all the quantities a_i, $i = \overline{1,\nu}$ and R_ν are regular functions on z.

The generalization of this results on the spin situation have been obtained in /3/.

The cluster estimates can be used also to establish the local limit theorems.

<u>Theorem 9.6.3 (/107/)</u>. Let $\Lambda_n \in \mathcal{A}(\nu)$, $n = 1, 2, \ldots$ $\lim\limits_{n \to \infty} |\Lambda_n| = \infty$ and let a potential Φ satisfy (9.6.1). Then for sufficiently small activity z

$$\Gamma^\Phi(J \in W : |J \cap \Lambda_n| = N) = \frac{1}{\sqrt{2\pi f_1(\Lambda_n, z)|\Lambda_n|}} \times$$

$$\times \exp\left\{\frac{(E|J \cap \Lambda_n| - N)^2}{2 f_1(\Lambda_n z)|\Lambda_n|}\right\}\left\{1 - \frac{f_2(\Lambda_n, z)}{6 f_1^{3/2}(\Lambda_n z)\sqrt{|\Lambda_n|}} \times\right.$$

$$\left. \times\left[\frac{3(E|J \cap \Lambda_n| - N)}{\sqrt{f_1(\Lambda_n, z)|\Lambda_n|}} - \left(\frac{E|J \cap \Lambda_n| - N}{\sqrt{f_1(\Lambda_n, z)|\Lambda_n|}}\right)^3\right] + O(|\Lambda_n|^{-1})\right\}$$

uniformly with respect to all N such that

$$|E|J \cap \Lambda_n| - N| < a\sqrt{|\Lambda_n|}, \quad a > 0.$$

Moreover

$$f_s(\Lambda_n, z) = (-i)^{s+1} \frac{d^{(s)} a_0}{d(z e^{it})^s}\Big|_{t=0} + (-i)^{s+1} \frac{d^{(s)} a_1}{d(z e^{it})^s}\Big|_{t=0} \times$$

$$\times \frac{1}{|\Lambda_n|} \sum_{\lambda \in \Lambda_n^{\nu-1}} |\lambda| + (-i)^{s+1} \frac{d^{(s)} R_1(\Lambda_n)}{d(z e^{it})^s}\Big|_{t=0}, \quad s = 1, 2,$$

where the notations have the same meaning as in the previous theorem.

Some other variants of the local limit theorems for Gibbs random fields are presented in /35/, /27/, /61/.

Note also the results concernig the c.l.th. for the energy (/82/, /61/,/85/) and some statistical applications, for example, the locally asymptotical normality of Gibbs random fields (/71/,/85/).

Some Additional Remarks

Concerning Gordin's method (S. 4.2) we additionally recommend the article of R.C.Bradley "On some results of M.I.Gordin:"A clarification of a misunderstanding" in Journal of Theoretical Probability, v.1, 2, 1988.

For a detailied explanation of Stein's method we address the reader to Louis H.Y. Chen's: "Stein's method in limit theorems for dependent random variables", Sea Bull.Math.(special issue), 1979.

In connection with moment's method we note the recently published book of L.Saulis and V.Statuliavičius "Limit theorems for large derivations", Vilnius, Mokslas Publishers, 1989. Here, in particular, the authors prove that the mixed cumulants of a random process can be estimated using mixing functions. For example, the following result holds:

<u>Theorem.</u> Let $X_1, X_2, ..., X_n$ be random variables and $\Gamma_k(S_n)$ be the k-order semi-invariant of $S_n = \sum_{k=1}^{n} X_k$. Let

$$\Lambda_n(\alpha, u) = max\left\{ 1, \max_{1 \leq s \leq u} \sum_{t=s} \alpha^{1/u}(s,t) \right\},$$

where $\alpha(s,t)$ is a α-mixing coefficient. Then

$$\left| \Gamma_k(S_n) \right| \leq 2k!\, 8^{k-1} c^k \Lambda_n^{k-1}(\alpha, k-1)\, n.$$

From this theorem via moments' method Theorem 5.1.7 of Ibragimov follows.

REFERENCES

1. ANSHELEVICH,V.V. (1973): The central limit theorem in non-
   commutative probability theory.-Dokl.Akad. Nauk SSSR 208,
   1265-1267.
2. ARZUMANIAN, V.A., NAHAPETIAN,B.S., POGOSIAN,S.K. (1986) :
   Cluster properties of classical lattice spin system. - Theor.
   Math. Phys. 67, 21-31.
3. ARZUMANIAN,V.A., NAHAPETIAN,B.S., POGOSIAN,S.K. (1988) :
   Asymptotic expansion of the logarithm of the partition function for the spin lattice systems. (to appear in Theor.Math.
   phys.).
4. AVERINCEV,M.A. (1970): On a method of describing discrete
   parameter random fields. Problemi Pered. Inform. 6, 100-109.
   (In Russian).
5. AVERINCEV, M.V. (1972) : The description of Markov random
   fields by Gibbs conditional distribution. Theory Probab.
   Appl. 17, 21-35.
6. BILLINGSLEY,P. (1961) : The Lindeberg-Levy theorem for mar-
   tingales. Proc.Amer. Math. Soc. 12, 788-792.
7. BILLINGSLEY,P. (1968): Convergence of probability measures.
   Wiley, New York.
8. BLUM,J.R., HANSON,D.L. and KOOPMANS,L.H. (1963): On the
   strong law of large numbers for a class of stochastic processes. Z. Wahrsch. Verw. Gebiete 2, 1-11.
9. BOLTHAUSEN, E. (1982): On the central limit theorem for stationary mixing random fields. Ann. Probab. 10, 1049-1052.
10. BRADLEY,R.C. (1980): A remark on the central limit question
    for dependent random variables. J.App.Prob. 17, 94-101.
11. BRADLEY,R.C.(1981): A sufficient condition for linear growth
    of variables in a stationary random sequence. Proc.Amer.
    Math.Soc. 83, 586-589.
12. BRADLEY,R.C. (1981): Central limit theorems under weak de-
    pendence. J. Multivariate Anal. 11, 1-16.
13. BRADLEY,R.C. (1983): Information regularity and central li-
    mit question. Rocky Mtn.J.Math. 13, 77-97.
14. BRADLEY,R.C. (1985): On the central limit question under ab-
    solute regularity. Ann. Prob. 13, 1314-1325.
15. BRADLEY,R.C. (1986): Basic properties of strong mixing con-
    ditions. Dependence in Probability and Statistics, ed.E.Eber-
    lein, M.Taqqu, Birkhäuser, Basel.

16. BRATTELI, O. and ROBINSON, D.W. (1979): Operator algebras and
    quantum statistical mechanics 1. Springer-Verlag,N.Y.,Heidelberg,
    Berlin.
17. BULINSKII,A.V. (1981): Limit theorems for random processes and
    fields. Moscow Univers.Press. (In Russian).
18. BULINSKII,A.V. (1986): An estimation of the convergence rate
    in the central limit theorem for random fields. Dokl.Akad.Nauk.
    SSSR, 291, 22-25.
19. BULINSKII,A.V. and ZURBENKO,I.G. (1976): Central limit theorem
    for additive random functions. Theory Probab.Appl.20, 707-717.
20. CHANDA,K.C. (1974): Strong mixing properties of linear stochas-
    tic processes. J.Appl. Probab. 11, 401-408.
21. CHUNG,K.L. (1960): Markov chains with stationary transition pro-
    babilities. Springer-Verlag, Berlin.
22. DAVYDOV,YU.A. (1968): Convergence of distributions generated by
    stationary stochastic processes. Theory Probab.Appl.13,691-696.
23. DAVYDOV,YU.A. (1969): On the strong mixing property for Markov
    chains with a countable number of states. Dokl.Akad.Nauk.SSSR,
    10, 825-827.
24. DAVYDOV,YU.A. (1970): The invariance principle for stationary
    processes. Theory Probab.Appl. 15, 487-498.
25. DAVYDOV,YU.A. (1973): Mixing conditions for Markov chains. Theo-
    ry Probab. Appl. 18, 312-328.
26. DEHLING,H., DENKER,M., PHILIPP,W. (1987): Central limit theorems
    for mixing sequences of random variables under minimal condi-
    tions. Ann. Probab.
27. DEL GROSSO,G. (1974): On the local central limit theorem for
    Gibbs processes. Commun. Math. Phys. 37, 141-160.
28. DENKER,M. (1986): Uniform integrability and the central limit
    theorem for strongly mixing processes. Dependence in Probabili-
    ty and Statistics, ed.E.Eberlein, M.Taqqu, Birkhäuser,Basel.
29. DOBRUSHIN,R.L. (1968): The discription of a random field by
    means of conditional probabilities and conditions of its regu-
    larity. Theory Probab. Appl. 13, 197-224.
30. DOBRUSHIN,R.L. (1968): Gibbs random fields for lattice systems
    with pair-wise interactions. Funct.Anal.Appl. 2, 292-301.
31. DOBRUSHIN, R.L. (1968): The problem of uniqueness of a Gibbsian
    random field and the problem of phase transitions. Funct.Anal.
    Appl. 2, 302-312.
32. DOBRUSHIN,R.L. (1970): Prescribing a system of random variables
    by conditional distributions. Theory Probab. Appl. 15, 458-486.

33. DOBRUSHIN, R.L. and MINLOS, R.A. (1967): Existence and continuity of pressure in classic statistical physics. Theory Probab. Appl. 12,

34. DOBRUSHIN,R.L. and NAHAPETIAN,B.S. (1974): Strong convexity of pressure for lattice systems of classical statistical physics. Theor. Math. Phys. 20, 223-234.

35. DOBRUSHIN,R.L. and TIROZZI,B. (1977): The central limit theorem and the problem of equivalence of ensembles. Commun. Math. Phys. 54, 173-192.

36. DOOB, J.L. (1953): Stochastic Processes, Wiley, N.Y.

37. DUNEAU,M., IAGOLNITZER,D. and SOUILLARD,B. (1975): Decay of correlations for infinite-range interactions. J.Math. Phys. 16,1662-1666.

38. DUNEAU,M. and SOUILLARD,B. (1976): Cluster properties of lattice and continuous systems. Commun. Math. Phys. 47, 155-166.

39. DVORETZKY,A. (1972): Asymptotic normality for sums of dependent random variables. Proc. 6th Berkeley Symp. on Math.Stat.and Probab.Univ. California Press, 2, 513-535.

40. GALLAVOTTI,G. and MIRACLE-SOLE,S. (1967): Statistical mechanics of lattice systems. Commun. Math. Phys. 5, 317-323.

41. GALLAVOTTI,G. and MIRACLE-SOLE,S. (1968): Correlation functions of a lattice system. Commun. Math. Phys. 7, 274-288.

42. GEBELEIN,H. (1941): Das statistische Problem der Korrelation als Variations- und Eigenwertproblem und sein Zusammenhang mit der Ausgleichungsrechnung. Z.Angew. Math.Mech. 21, 364-379.

43. GOLDIE, C.M., MORROW,G.J. (1986): Central limit questions for random fields. Dependence in Probability and Statistics, ed.E. Eberlein, M.Taqqu. Birkhauser, Basel.

44. GOLDIE,M. and GREENWOOD,P. (1986): Characterisations of set-indexed Brownian motion and associated conditions for finite-dimensional convergence. Ann. Probab. 14, 802-816.

45. GOLDIE,M. and GREENWOOD,P. (1986): Variance of set-indexed sums of mixing random variables and weak convergence of set-indexed processes. Ann. Probab. 14, 817-839.

46. GOLDSTEIN,V.G. (1982): A central limit theorem of "non-commutative" probability theory. Theory Probab. Appl. 27, 657-666.

47. GORDIN, M.I. (1969): The central limit theorem for stationary processes. Dokl.Akad.Nauk.SSSR 10, 1174-1176.

48. GORDIN,M.I. (1971): On the behaviour of the variance of sums of random variables that form a stationary process. Theory Probab. Appl. 16, 484-494.

49. GORODETSKII, V.V. (1977): On the strong mixing property of li-
    nearly generated sequences. Theory Probab.Appl. 22, 421-423.
50. GORODETSKII, V.V. (1982): The invariance principle for stationa-
    ry random fields, satisfying the strong mixing condition. Theory
    Probab. Appl. 27, 358-364.
51. GORODETSKII, V.V. (1984): The central limit theorem and an in-
    variance principle for weakly dependent random fields. Dokl.
    Akad. Nauk SSSR, 276, 528-531.
52. GORODETSKII, V.V. (1985): Moment inequalities and the central
    limit theorem for integrals of mixing random fields. Zap.Nauchn.
    Sem. LOMI, 142, 39-47.
53. GRIFFITHS,R. and RUELLE,D. (1971): Strict convexity (Continuity)
    of the pressure in lattice systems. Commun.Math.Phys.23,169-175.
54. GROSS, L. (1979): Decay of correlations in classical lattice
    models at high temperature. Commun.Math.Phys.68, 9-27.
55. HALFINA, A.M. (1969): The limiting equivalence of the canonical
    and grand canonical ensembles (low density case). Math. Sbornik
    80, 3-51. (In Russian).
56. HALL,P. and HEYDE C.C. (1980): Martingale limit theory and its
    application. Academic-Press, New-York.
57. HEGERFELDT, G.C. and NAPPI,C.R. (1977): Mixing properties in
    lattice systems. Commun. Math. Phys. 53, 1-7.
58. HERRNDORF,N. (1983): Stationary strongly mixing sequences not
    satisfying the central limit theorem. Ann. Prob. 11, 809-813.
59. HERRNDORF, N. (1984): A functional central limit theorem for
    weakly dependent sequences of random variables. Ann. Probab.12,
    141-153.
60. HEYDE, C.C. (1974): On the central limit theorem for stationary
    processes. Z.Wahrsch.Verw.Gebiete 30, 315-320.
61. IAGOLNITZER,D. and SOUILLARD B. (1979): Random fields and limit
    theorems. Random Fields (eds. Fritz,J., Lebowitz,J.I.,Szasz,D.)
    Colloquia mathematica societatis Janos Bolyai, 27.
62. IBRAGIMOV, I.A. (1959): Some limit theorems for strictly statio-
    nary stochastic processes. Dokl.Akad.Nauk SSSR 125, 711-714.
    (In Russian).
63. IBRAGIMOV,I.A. (1962): Some limit theorems for stationary pro-
    cesses. Theory Probab. Appl. 7, 349-382.
64. IBRAGIMOV, I.A. (1963): A central limit theorem for a class of
    dependent random variables. Theory Probab. Appl. 8, 83-89.
65. IBRAGIMOV, I.A. (1975): A note on the central limit theorem for
    dependent random variables. Theory Probab. Appl. 20, 135-141.

66. IBRAGIMOV, I.A.; LINNIK, Y.V. (1971): Independent and stationary sequences of random variables. Wolters-Noordhoff, Groningen.
67. IBRAGIMOV, I.A. and ROZANOV, Y.A. (1978): Gaussian random processes. Nauka, Moscow. (In Russian).
68. IOSIFESCU, M. (1968): The law of the iterated logarithm for a class of dependent random variables. Theory Probab. Appl. 13, 304-313.
69. IOSIFESCU,M. (1980): Recent advances in mixing sequences of random variables. Third International Summer School on Probability Theory and Mathematical Statistics, Varna, 1978.
70. IOSIFESCU,M. and TEODORESCU,R. (1969): Random processes and learning. Springer-Verlag, N. Y.
71. JANSURA, M. (1985): Central limit theorems for random fields with application to locally asymptotic normality of Gibbs random fields (Preprint).
72. KANTOROVICH,L.V. and RUBINSTEIN,G.S. (1958): On a space of completely additive Functions. Vestnik Leningr.Gos.Univ. 7, 52-59. (In Russian).
73. KESTEN,H. and O'BRIEN,G.L. (1976): Examples of mixing sequences. Duke Math. J. 43, 405-415.
74. KOLMOGOROV, A.N. and ROZANOV, Y.A. (1960): On strong mixing conditions for stationary Gaussian processes. Theory Probab. Appl. 5, 204-208.
75. KÜNSCH,H. (1982): Decay of correlation under Dobrushin's uniqueness condition and its applications. Commun. Math. Phys. 84, 207-222.
76. LANFORD,O.E. and RUELLE,D. (1969): Observables at infinity and states with short range correlations in statistical mechanics. Commun. Math. Phys. 13, 194-215.
77. LEONENKO,N.N. and JADRENKO,M.I. (1979): On the invariance principle for homogeneous isotropic random fields. Theory Probab. Appl. 24, 175-181.
78. LEONOV, V.P. (1961): On the dispersion of time-dependent means stationary stochastic process. Theory Probab.Appl.6, 93-101.
79. LOEVE, M. (1960): Probability theory. D.Van Nostrand.
80. MALYSHEV,V.A. (1975): The central limit theorem for Gibbsian random fields. Dokl. Akad. Nauk SSSR 16, 1141-1154.
81. MALYSHEV,V.A., MINLOS, R.A. (1985): Gibbsian fields: the method of cluster expansions. Moscow, Nauka. (In Russian).
82. MINLOS,R.A. and HALFINA,A.M. (1970): Central limit theorem for the energy and the number of particles in lattice systems of gas. Izv.Akad.Nauk SSSR, 34, 1173-1191 (In Russian).

83. NAHAPETIAN,B.S. (1975): Strong mixing condition for Gibbs random fields with discrete parameter and some applcations. Izvestia Akad. Nauk Arm.SSR X, 242-254.

84. NAHAPETIAN,B.S. (1975): Central limit theorem for random fields satisfying strong mixing condition. Dokl.Akad.Nauk Arm.SSR 61, 210-213.

85. NAHAPETIAN,B.S. (1979): Limit theorems and statistical mechanics. Random Fields (ed. Fritz,J., Lebowitz,J.I., Szasz,D.)Colloquia mathematica societatis Janos Bolyai, 27.

86. NAHAPETIAN,B.S. (1980): On a criterion of weak dependence. Theory Probab. Appl. 25, 374-381.

87. NAHAPETIAN,B.S. (1980): The central limit theorem for random fields with mixing property. Multicomponent random systems, Dobrushin,R.L., Sinai,Ya.G. (eds.). Dekker, New York, Basel.

88. NAHAPETIAN,B.S. (1984): On limit theorems for dependent random variables. The 6th International Symposium on Information Theory. Tashkent.

89. NAHAPETIAN,B.S. (1986): On limit theorems for dependent random variables. Dokl.Akad.Nauk Arm.SSR 82, 99-101.

90. NAHAPETIAN,B.S. (1987): An approach to limit theorems for dependent random variables. Theory Probab.Appl. 32, 589-595.

91. NAHAPETIAN,B.S. (1988): On the Dobrushin's hypothesis. (to appear in Probab. and Math.Statist., Wroclaw).

92. NEADERHOUSER,C.C. (1978): Limit theorems for multiply indexed mixing random variables, with application to Gibbs random fields. Ann. Probab. 6, 207-215.

93. NEADERHOUSER,C.C. (1978): Some limit theorems for random fields. Commun. Math. Phys. 61, 293-305.

94. NEADERHOUSER,C.C. (1980): Convergence of block spins defined by a random field. J. Statist. Phys. 22, 673-683.

95. NEWMAN,C.M. (1980): Normal fluctuations and the FKG inequalities. Commun. Math. Phys. 74, 119-128..

96. OODAIRA,H. and YOSHIHARA,K. (1971): The law of the iterated logarithm for stationary processes satisfying mixing conditions. Kodai Math. Sem. Rep. 23, 311-334.

97. OODAIRA,H. and YOSHIHARA,K. (1972): Functional Central limit theorems for strictly stationary processes satisfying the strong mixing condition. Kodai Math. Sem. Rep. 24, 259-269.

98. PARTHASARATHY, K.R. (1967): Probability measures on metric space. Academic Press, N. Y.

99. PELIGRAD, M. (1982): Invariance principles for mixing sequences of random variables. Ann. Probab. 10, 968-981.

100. PELIGRAD,M. (1985): An invariance principle for φ -mixing
     sequences. Ann. Prob. 13, 4, 1304-1313.
101. PELIGRAD,M. (1986): Recent advances in the central limit theo-
     rem and its invariance principle for mixing sequences of random
     variables. Dependence in Probability and Statistics, ed.E.Eber-
     lein, M.Taqqu. Birkhauser, Basel.
102. PETROV,V.V. (1972): Sums of independent random variables. Nauka,
     Moscow. (In Russian).
103. PHILIPP, W. (1986): Invariance principles for independent and
     weakly dependent random variables. Dependence in Probability
     and Statistics, ed. E.Eberlein,M.Taqqu. Birkhauser,Basel.
104. PHILIPP,W. and STOUT,F. (1975): Almost sure invariance princi-
     ples for partial sums of weakly dependent random variables.
     Mem. Amer. Math. Soc. 161.
105. PINSKER,M.S. (1964): Information and information stability of
     random variables and processes. Holden-Day, San Francisco.
106. PRESTON.C (1976): Random fields. Lecture Notes in Mathematics
     534. Berlin, Heidelberg, New York, Springer.
107. POGOSIAN,S. (1979): Asymptotic expansion in the local limit
     theorem for the particle number in the grand canonical ensem-
     ble. Random Fields (eds. Fritz,J., Lebowitz,J.I., Szasz,D.)
     Colloquia mathematica societatis Janos Bolyai, 27.
108. POGOSIAN,S. (1984): Asymptotic expansion of the logarithm of
     the partition function. Commun. Math. Phys. 95, 227-245.
109. RESNIK, M.H. (1968): The law of the iterated logarithm for some
     classes of stationary processes. Theory Probab. Appl. 13, 642-
     656.
110. RJAUBA,B. (1962): The central limit theorem for sums of series
     of weakly dependent random variables. Litovsk.Math.Sb.2,193-205.
111. ROSENBLATT,M. (1956): A central limit theorem and a strong
     mixing condition. Proc.Nat.Acad.Sci.U.S.A. 42, 43-47.
112. ROSENBLATT,M. (1971): Markov processes, structure and asympto-
     tic behaviour. Springer-Verlag, Berlin.
113. ROZANOV,Y.A. (1963): Stationary random process. Nauka, Moscow.
114. RUELLE,D. (1969): Statistical mechanics. Benjamin, New York.
115. SARIMSAKOV,T.A. (1985): An introduction to quantum probability
     theory. Fan, Tashkent.
116. SINAI,YA.G. (1962): On limit theorems for stationary processes.
     Theory Probab. Appl. 7, 213-219.
117. SPITZER,F. (1971): Markov random fields and Gibbs ensembles.
     Amer.Math.Monthly, 78, 142-154.

118. STATULEVICUS,V.A. (1970): On limit theorems for random functions I. Liet. Math.sbornik. 10, 583-592. (In Russian);
119. STATULEVICIUS, V.A. (1974): Limit theorems for dependent random variables under various regularity conditions. Proc.Int. Congr. Math.Vancouver. 2, 173-181.
120. STEIN, CH. (1973): A bound for the error in the normal approximation of a sum of dependent random variables. Proc. 6th Berkley Symp. Math.Stat. and Prob. 2, 583-602.
121. SUNKLODAS, J. (1986): An estimation of the convergence rate in the central limit theorem for weakly dependent random fields. Litovsk. Math.Sb. 26, 541-559.
122. TAKAHATA,H. (1982): On the central limit theorem for weakly dependent random fields. Yokohama Math. J., 31, 67-77.
123. TIKHOMIROV,A.N. (1980): On the rate of convergence in the central limit theorem for weakly dependent random variables. Theory Probab. Appl. 4, 800-818.
124. TROTTER, H. (1959): On the product of semigroups of operators. Proc. Amer. Math. Soc. 10, 545-551.
125. VALLANDER, S.S. (1973): Calculation of the Waserstein distance between probability distributions on the line. Theory Probab. Appl. 18, p. 824-827.
126. VASERSHTEIN, L.N. (1969): Markov processes over denumerable products of spaces, describing large systems of automata. Probl. Inform. Transm. 5, 64-72.
127. VOLKONSKII, V.A. and Y.A.ROZANOV (1959): Some limit theorems for random function I. Theory Probab. Appl. 4, 178-197.
128. VOLKONSKII, V.A. and ROZANOV,Y.A. (1961): Some limit theorems for random function II. Theory Probab.Appl. 5, 186-198.
129. WITHERS, C.S. (1981); Central limit theorems for dependent random variables I. Z.Wahrsch.Verw.Gebiete 57, 509-534.
130. YOSHIHARA,K. (1978): Probability inequalities for sums of absolutely regular processes and their applications. Z.Wahrsch. verw.Gebiete, 43, 319-329.
131. YOKOYAMA,R. (1980): Moment bounds for stationary mixing sequences. Z. Wahrsch. verw. Gebiete 52, 45-57.
132. ZURBENKO, I.G. (1987): Analysis of stationary and homogeneous random systems. Moscow Univers. Press. (In Russian).

ADDITION.

133. BULINSKII,A.V. (1986): The central limit theorem and invariance principle for mixing random fields. Abstracts of communi-

cations of 1-st World congress of Bernoulli Society,USSR,Tashkent,640.

134. GNEDENKO,B.V.,KOLMOGOROV,A.N.(1948): Limit distributions for sums of independent random variables.Moscow.(In Russian).

135. GUYON,X.,RICHARDSON,S.(1984): Vitesse de convergence du theoreme de la limite centrale pour des champs faiblement dependants. Z. Wahrsch. verw. Gebiete,66,297-314.

136. HEINRICH,L. (1982):A method for the derivation of limit theorems for sums of m-dependent random variables. Z. Wahrsch. verw. Gebiete,60,501-515.

137. HEINRICH,L.(1987): Rates of convergence in stable limit theorems for sums of exponentially Ψ-mixing random variables with an application to the metric theory of continued fractions. Math. Nachr. 131,149-165.

138. HEINRICH,L. (1990): Limit theorems for m-dependent random variables and their applications.Teubner-Verlag,Leipzig (in preparation).

139. SENETA ,E. (1976): Regularly varying functions.Lecture Notes in Mathematics, 508. Springer-Verlag, Berlin-Heidelberg-New York.

140. SERFLING,R.J. (1970): Moments inequalities for the maximum cumulative sum. Ann. Math. Statist.,41,1227-1234.

141. UTEV,S.A. (1984): Inequalities for sums of dependent random variables and the rate of convergence in invariance principle. In "Limit theorems for sums of random variables". Novosibirsk.